Christian Mennerich

Phase-field modeling of multi-domain evolution in ferromagnetic shape memory alloys and of polycrystalline thin film growth

**Schriftenreihe
des Instituts für Angewandte Materialien**
Band 19

Karlsruher Institut für Technologie (KIT)
Institut für Angewandte Materialien (IAM)

Phase-field modeling of multi-domain evolution in ferromagnetic shape memory alloys and of polycrystalline thin film growth

by
Christian Mennerich

Dissertation, Karlsruher Institut für Technologie (KIT)
Fakultät für Maschinenbau
Tag der mündlichen Prüfung: 30. Januar 2013

Impressum

Karlsruher Institut für Technologie (KIT)
KIT Scientific Publishing
Straße am Forum 2
D-76131 Karlsruhe
www.ksp.kit.edu

KIT – Universität des Landes Baden-Württemberg und
nationales Forschungszentrum in der Helmholtz-Gemeinschaft

Diese Veröffentlichung ist im Internet unter folgender Creative Commons-Lizenz
publiziert: http://creativecommons.org/licenses/by-nc-nd/3.0/de/

KIT Scientific Publishing 2013
Print on Demand

ISSN 2192-9963
ISBN 978-3-7315-0009-4

Zusammenfassung

Die Phasenfeldmethode ist ein mächtiges Werkzeug für die compterge-stützte Analyse der Evolution von Mikrostrukturen auf mesoskopischen Längenskalen in Raum und Zeit. Sie wird häufig in den Materialwissenschaften angewandt. In dieser Arbeit wird ein Phasenfeldmodell, das mit Hilfe von unterschiedlich räumlich orientierten skalarwertigen Ordnungsparametern komplexe Mikrostrukturgeometrien abbilden kann, verwendet, um Computersimulationen in verschiedenen materialwissenschaftlich relevanten Gebieten durchzuführen. Das Hauptaugenmerk liegt auf der Entwicklung numerischer Methoden zur Kopplung der Evolution der Ordnungsparameter mit der Evolution langreichweitiger Felder wie der elastischen Dehnung oder der spontanen Magnetisierung. Dies ermöglicht die Beschreibung der Evolution von Multidomänenstrukturen in Mikrostrukturen, die aus kristallographischen und magnetischen Domänen bestehen. Techniken und Randbedingungen, die es erlauben das finite Rechengebiet als representatives Volumen zu betrachten, werden angewendet, was die Beschränkung auf periodische Mikrostrukturen erlaubt. Die Annahme der Periodizität macht die Anwendung von FFT-Techniken möglich. Die Evolution der spontanen Magnetisierung unterliegt geometrischen Zwängen, weshalb eine Integrationsmethode diskutiert und implementiert wird, die diese Bedingungen auf natürliche Weise erfült. Das Phasenfeldmodell wird auf zwei wichtige Modellsysteme angewandt, die beide sowohl von einem materialwissenschaftlichen als auch einem industriellen Standpunkt aus von Interesse sind.

Das Erste ist das anisotrope Wachstum von Kristalliten auf einem glatten Substrat in eine hydrothermale Lösung. Die Ordnungsparameter des Phasenfeldmodells werden zur Beschreibung der Orientierung der Kristallite verwendet. Das Wachstumsverhalten wird eingehend am Beispiel des konkurrierenden Wachstums Zeolith-artiger Kristalle bei der Bildung dünner Filme untersucht. Zeolithkristalle dienen hier als Modellsystem zur

Analyse von durch Anisotropien in der Oberfächenenergie oder in der Anlagerungskinetik getriebenen Systemen. Zeolithische dünne Filme dienen in der Ölindustrie als Katalysatoren und molekulare Siebe.

Die zweite Anwendung des entwickelten Phasenfeldmodells ist die Untersuchung von magnetischen Formgedächtnismaterialien, einer Klasse aktiver Materialien, die als Komponenten in Aktuatoren und Dämpfern Verwendung finden. Magnetische Formgedächtnismaterialien erlauben große makroskopische Dehnungen, die durch die Reorganisation einer martensitischen Mikrostruktur hervorgerufen werden, indem an das Material von Außen ein Magnetfeld angelegt wird. Notwendige Voraussetzung für das Auftreten des magnetischen Formgedächtniseffekts ist eine vorhergehende martensitische Transformation, die das Material aus einer austenitischen Elternphase in eine martensitische Mikrostruktur überführt, die aus unterschiedlich orientierten, aber kristallographisch und energetisch äquivalenten Martensitvarianten besteht. Auf der mesoskopischen Längenskala basiert der magnetische Formgedächtniseffekt auf einer komplizierten
Wechselwirkung zwischen elastischen und ferromagnetischen Domänen. Die elastischen Domänen hängen zusammen mit den Eigendehnungen, die durch die martensitische Transformtion hervorgerufen wurden. Die ferromagnetischen Domänen sind an magnetisch bevorzugte kristallographische Richtungen in den martensitischen Varianten gebunden. Das entwickelte Phasenfeldmodell findet Verwendung zur Untersuchung des magnetischen Formgedächtniseffekts und verwandter Phänomene wie magnetischer Hysterese und Spannungs-Dehnungs-Beziehungen in der Heuslerlegierung Ni_2MnGa, einem eingehend untersuchten magnetischen Formgedächtnismaterial, das hier als Modellsystem dient.

Abstract

The phase-field method is a powerful tool to be used for computer-aided analysis of the time-spatial evolution of materials' microstructures on the mesoscale, and is often applied in materials science. In this work, a multi-phase-field model that is capable of treating various differently oriented scalar order parameters to describe complex microstructure geometries will be adopted to run computer simulations in different areas of materials science. The main focus will be on the development of numerical methods to couple the evolution of the order parameters to the evolution of long-range fields like elastic strain or spontaneous magnetization. This allows for simulations of the multi-domain evolution in microstructures that consist of simultaneously evolving crystallographic and ferromagnetic domains. Techniques and boundary conditions that treat the computation domain as a representative volume element will be applied that allow to restrict considerations to periodic microstructures. The assumption of periodicity makes FFT-techniques applicable. To account for geometric constraints that arise in micromagnetism, an integration method will be discussed and implemented that is unconditionally norm conservative. The phase-field model will be applied to two important cases of high scientific and industrial interest.

The first is the anisotropic concurrent growth of crystallites on a smooth substrate into a hydrothermal solution that will be studied on the example of the growth of zeolite-like crystals that constitute thin films. The order parameters of the phase-field model represent the different growth directions of the crystallites. Besides being an interesting model system for the study of growth competition driven by surface energy anisotropy and kinetic anisotropy, zeolite thin films have an important application in oil industry as catalytic active supports or molecular sieves.

The second application of the developed model is the analysis of the behavior of magnetic shape memory alloys, a class of active smart martensitic materials that are used as components in actuators and dampers. Magnetic shape memory alloys provide giant macroscopic strains caused by the rearrangement of the microstructure by application of an external magnetic field. A preceding martensitic transformation is a necessary requirement, so the materials that offer the magnetic shape memory effect consist of differently oriented but crystallographically and energetically equivalent martensitic variants that arise as deformations from a common austenite parent phase. On the mesoscale, the magnetic shape memory effect is based on the complex interplay of elastic and ferromagnetic domains. The elastic domains are determined by the eigenstrains of the martensitic variants that arise from the transformation from the parent phase. The ferromagnetic domain structure is linked to crystallographic direction in the martensitic variants that serve as magnetically preferred directions. The developed phase-field model will be used to investigate the magnetic shape memory effect and related phenomena like magnetization hysteresis and external stress vs. strain behavior in the Heusler alloy Ni_2MnGa, a well studied magnetic shape memory alloy model system.

Acknowledgments

Many people have been involved in the development of this work. The people of the working group of Prof. Nestler at the Karlsruhe University of Applied Sciences and the Karlsruhe Institute of Technology in which I was integrated during the time I worked on this thesis have provided an interdisciplinary discussion forum. Some of my colleagues have become friends over the years.

I thank Prof. Dr. Britta Nestler for her guidance and her support throughout the time I worked on this thesis. Prof. Nestler has encouraged me from the early stages of my work to discuss it with people from other groups at conferences and meetings.

Dr. Frank Wendler has been closely involved in many parts of my work in the group. He has taught me a lot about the physical interpretation of numerical simulations. He has always had an open ear for questions and has been willing to discuss problems through again and again.

Marcus Jainta helped a lot with his detailed knowledge of the software architecture used in the group. He has done much of the programming work, run a lot of simulations and contributed to the analysis of the simulation data.

Dr. Anastasia August has been involved in discussions about the mathematical basics related to (Lie-)group theory that are presented in the first section of this work. She also has provided help on some of the examples presented there.

I acknowledge the help all members of the scientific work group of Prof. Nestler have provided over the years, actual and former members alike. I especially acknowledge Anna Roes, who has realized in great parts the implementation of the solution method for a mechanical equilibrium within her Karl-Steinbuch Scholarship, and Dr. A. Melcher.

I thank my wife Laurence for her constant support and patience over the last years while I was working on this thesis. She critically read this text and commented on the language and grammar I used in its early versions.

I gratefully acknowledge the financial support that was granted: The funding of the Center of Computational Materials Science and Engineering (CCMSE) through the Landesstiftung Baden-Württemberg and the European Fonds EFRE, and the funding of the German Research Foundation (DFG) within the priority programme SPP 1239 ('Change of microstructure and shape of solid materials by external magnetic fields').

Contents

IV Application and Outlook   151

9 A phase-field model for polycrystalline thin film growth   153

10 Phase-field modeling of the magnetic shape memory effect   181

11 Outlook   201

V Appendix   209

A Interpretation and representation of rotations   211

B The Voigt notation for elasticity   227

List of Figures

List of Tables

Symbols and Abbreviations

$\forall$	for all
$\exists$	there exists
$\exists!$	there exists exactly one
$\leftrightarrow$	equivalency
$\rightarrow$	implication
$\in$	element-of relation in set theory
$\subseteq$	subset relation
$\subset$	proper subset relation
$\trianglelefteq$	normal subgroup relation
$\ltimes$	semi-direct product of groups
$\mathbb{N}$	set of natural numbers including 0
$\mathbb{Z}$	set of integer numbers
$\mathbb{R}$	field of real numbers
$\mathbb{R}_{>0}$	set of real numbers greater than zero
$\mathbb{R}_{\geq 0}$	set of real numbers greater than or equal to zero
∞	symbol for (positive) infinity
$\mathbb{C}$	field of complex numbers

i	imaginary complex unit, $i = \sqrt{-1}$
$\mathbb{K}^n$	n-dimensional vector space over a field $\mathbb{K}$
$\mathbf{0}$	zero-vector of $\mathbb{K}^n$
$\mathbb{H}$	skew field of quaternions
$\mathbb{S}^{n-1}$	unit sphere in $\mathbb{R}^n$
$\mathrm{SO}(n)$	Lie-group of rotations in $\mathbb{R}^n$
$\mathfrak{so}(n)$	Lie-algebra to $\mathrm{SO}(n)$
$\hat{f}$	Fourier transform of a function $f : \mathbb{C} \to \mathbb{C}$
$\boldsymbol{\sigma}$	elastic stress
$\boldsymbol{\epsilon}$	elastic strain
$\boldsymbol{\epsilon_0}$	eigenstrain
$\mathcal{C}$	elastic stiffness tensor
$\mathrm{E}_{\mathrm{elast}}$	elastic energy
f_{elast}	elastic energy density
κ	elastic damping parameter
M_S	saturation magnetization
μ_0	permeability in the vacuum
A_{exch}	magnetic exchange constant
K_{aniso}	uniaxial magnetocrystalline anisotropy constant
f_{ext}	magnetic Zeeman energy density
f_{demag}	demagnetization energy density

f_{exch}	magnetic exchange energy density
f_{aniso}	magnetic anisotropy energy density
$f_{\mathrm{m\text{-}el}}$	magneto-elastic energy density
$\mathrm{E}_{\mathrm{magnetic}}$	magnetic energy
f_{magnetic}	magnetic energy density
$\mathbf{p}$	direction of magnetic easy axis
$\mathbf{H}_{\mathrm{ext}}$	external magnetic field
$\mathbf{H}_{\mathrm{demag}}$	demagnetization field
$\mathbf{H}_{\mathrm{exch}}$	magnetic exchange field
$\mathbf{H}_{\mathrm{aniso}}$	magnetic anisotropy field
$\mathbf{H}_{\mathrm{m\text{-}el}}$	magneto-elastic field
$\mathbf{H}_{\mathrm{eff}}$	effective magnetic field
γ	gyromagnetic ratio
α_G	Gilbert damping parameter
ϕ	phase-field vector
ϕ_α	phase field α
$h(\phi)$	phase field interpolation function h
$\gamma_{\alpha\beta}$	α/β surface tension
$\mathcal{G}$	Gibbs simplex
τ	interface relaxation parameter
λ	Lagrange multiplier in the phase-field equation

f_{liquid}	bulk free energy of the liquid phase
f_{solid}	bulk free energy of the solid phase
FFT	Fast Fourier transform
MSMA	magnetic shape memory alloy
MSME	magnetic shape memory effect
MT	martensitic transformation
RVE	representative volume element
SOR	sucessive over-relaxation

Part N

Introduction

1 Introduction

The work at hand presents results to which the author has contributed as a member of the scientific group of Prof. Nestler since October 2008. Prof. Nestler's group is settled both at the Karlsruhe University of Applied Sciences - Institute of Materials and Processes (IMP), and the Institute of Applied Materials - Institute of Reliability of Systems and Components at the Karlsruhe Institute of Technology (KIT IAM - ZBS). In the beginning it was not clear what direction the work in the group would take. The general task was set out to be the development of a phase-field model to simulate the evolution of microstructures under elastic and magnetic forces. This requires knowledge on the fields of materials sciences and materials modeling, the multi phase-field method to be applied (developed by Nestler et al. [1] and well established), the numerical challenges of modeling mechanically and especially micromagnetically driven solid-to-solid phase transitions, and the real material systems the modeling process is all about.

The first step towards understanding the phase-field method was to simulate and analyze the competitive growth of MFI zeolite-like grains on thin films, a scenario that has interesting scientific and industrial applications. This study had been initiated by discussions with Prof. P.D. Bons and Dr. J. Becker (both from the University of Tübingen at that time), with whom the group was still in contact after former collaboration (see the article [2] for an example). The simulation studies that had been carried out resulted in an article that was published in 2011 in the *Journal of Crystal Growth* [3], and in great parts supervised and motivated by Dr. Frank Wendler. In parallel, the group of Prof. Nestler participated in the Priority Programme 1239 of the German Research Foundation (DFG SPP 1239: 'Change of microstructure and shape of solid materials by external magnetic fields'[1]). The goal envisaged here was to develop a phase-field

---

[1]See http://www.magneticshape.de/.

model to describe the rearrangement of a martensitic microstructure due to the application of an external magnetic or elastic stress field. The work in this field resulted in two major publications [4, 5] so far. The latter is benevolently cited in the literature (see e.g. [6, 7]). A third publication that discusses more recent results (which are also presented in this work) has been accepted for publication in the European Physical Journal B and will be published in the first quarter of 2013 in a special issue called *New trends in magnetism and magnetic materials* [8]. The author has been invited to submit this article by the organization committee of the Joint European Magnetic Symposia (JEMS) 2012 in Parma (Italy), where he had presented the phase-field modeling approach for magnetic shape memory alloys. The author tried to combine two tasks that are very different at first glance, but have in common the modeling approach on which they are based: The phase-field modeling of the competitive growth of grains on thin films and of effects related to the microstructure rearrangement in magnetic shape memory materials.

The rest of this introduction is split into two parts: The first is motivating the actual form of the text by explaining the aim behind the way it is written, and the second comments on the structure of the text and points out what might be left out by readers who know about the topics specific chapters deal with.

1.1 Motivation

As the author had little background on materials modeling when he began working in the group of Prof. Nestler, and because there was at that time little background in the group on the modeling of coupled elastic and micromagnetic processes, instructive material was gathered, mostly textbook material and review articles of the respective fields. Especially the numerical simulation of micromagnetic processes is very challenging because of constraints that have to be maintained during time integration. Here, numerically accurate but at the same time efficient solution methods are crucial. While starting to learn about new topics and problems by reviewing the different approaches already published in the literature, it was sometimes hard to recognize the promising ones that in addition would match the scientific environment the author was embedded into. So, the idea developed to find a way to adequately write down everything that enters the process of phase-field modeling for the given problems. In a simplified view, this involves the following steps:

1. Understanding the physical problems.
2. Describing the physical problems in an adequate mathematical language.
3. Constructing a mathematical model.
4. Solving the governing equations numerically.

The first two steps are independent of the last two, but it is necessary that the model and its numerical implementation reflect the properties that can be derived from the first two.

As learning is often a 'top down' process, the way solutions are presented is usually 'bottom up'. This philosophy is followed here, too, with the aim to write down and define all structures and properties that enter the processes of understanding the problem, of describing it adequately in a mathematical language, of developing the model and of solving the governing equations in such a way that all necessary constraints are met. Articles and even textbooks often take fundamental things for granted[2] and sometimes leave out (or seem not even to be aware of) difficulties coming along with the description of certain problems in the special language of mathematics. It should exemplary be shown how many assumptions are needed when it comes to modeling material processes, and how many of them are intuitively made or silently and implicitly taken for granted. It soon became clear that this idea could not be realized without writing a textbook, which was far out of scope of this thesis. The knowledge and fundamental mathematics necessary to properly handle the problems arising in this work, their description and adequate numerical solution as well, are gathered in the first part of this work. All the topics discussed there enter the modeling process in one way or the other. A few examples shall be given here:

---

[2]such as a certain intuitive knowledge of vector space algebra, of group theory, of differential equations etc

- The *theory of finite groups* is needed to mathematically capture the physical concept of a crystal structure. It enters the definition of crystal symmetries, and as a consequence, the way the number of different martensitic variants in magnetic shape memory materials can be counted.

- The *theory of Lie-groups* (or *continuous groups*) is needed to understand a solution method that fulfills the geometric constraint for the micromagnetic evolution equations without explicit projection of possible solutions onto the space of allowed solutions.[3]

- *Vector spaces* appear almost everywhere, as many important physical structures carry a vector space structure. Not only the ambient space is an example, but also special sets of matrices that serve as representations of symmetry operations.

The same language that is used to characterize symmetries and continuous transformations can be used to characterize real world structures like crystals and microstructures. Hence, the text starts from algebraic group and Lie-group theory, moves on to the definition and basic properties of vector spaces, basic analysis and Fourier transforms. Though the intended strategy could not be followed through in all parts of the text, it is the author's hope to motivate two things: First, that many theories enter the process of modeling material behavior. Sometimes some of these are accepted tacitly, which can cause confusion and unexpected problems. And second, that it is sometimes useful to look at mathematical theories, even if they might appear somewhat cumbersome at a first glance. Physics and mathematics might influence each other positively when it comes to the mathematical description of a model for physical processes and the solution of the resulting equations.[4] On understanding special theories one

---

[3]This scheme is more adequate than projection schemes, as the latter can alter other physical properties of the system. This is discussed in more detail in Chap. 8.

[4]An example that will not be treated in this work, but fits into the context of modeling the magnetic shape memory effect and the aforegoing martensitic transformation, is the field of so called Γ-convergence (see e.g. the script provided by A. Braides [9] on the theory, its application to the martensitic transformation described in the book of Bhattacharya [10], and the transfer to magnetic shape memory alloys by de Simone and James [11]).

might gain elegant solution methods, and also generalizations to other problems.[5]

1.2 Organization of the text

The first part of this text introduces the mathematical basics used to describe the physical processes modeled in this work (anisotropic grain growth and rearrangement of martensitic microstructures by external mechanic and magnetic fields). This first part might be skipped by readers who are solely interested in the modeling process or the simulation results, as many things presented here can be intuitively taken for granted.

The second part deals with the modeling of material behavior and the underlying physical theories. This part is more concrete than the first one, but uses in many parts the same mathematical language. Two continuum theories, continuum mechanics and micromagnetism, will be briefly discussed. They enable the description of the functional principles of a class of active and smart materials: Ferromagnetic shape memory alloys.

The third part introduces the phase-field method. First, some general aspects of this modeling approach will be discussed and then, the special model that serves as a basis for the modeling approaches of this work will be introduced. At this point a few words on the so called Landau theory will be given to distinguish this phase-field model from others that are published in the literature. Landau theory has been applied by other groups to develop phase-field models for solid-to-solid phase transitions. The chosen model has been numerically implemented and integrated into the software environment Pace3D[6], a software framework written in the programming language C that has been developed and maintained in the work group of Prof. Nestler for many years. The finite differences method and the implemented explicit integration schemes will be briefly reviewed, and new techniques and boundary conditions will be discussed. A special

---

[5]Again, Lie-group theory shall serve as an example. This theory might be considered to be neither easy nor intuitive, but it might provide interesting solution methods not only for the integration problem discussed in this work, but also for other problems like the description of the motion of rigid bodies [12].

[6]<u>P</u>arallel <u>a</u>lgorithms for <u>c</u>rystal <u>e</u>volution in <u>3D</u>.

algorithm to solve for the mechanical equilibrium during the microstructure evolution will be proposed. The numerical methods for the micromagnetic evolution equation and the computation of the demagnetization field require special solution techniques and will be presented in a separate chapter. The numerical accurate computations make extensive use of the mathematical theories discussed in the first part.

In the fourth part of this work, the developed models will be applied to simulations and analysis in two scientific and industrial interesting fields: First, the competitive growth of zeolite-like grains on thin films, which are used to grow molecular sieves for fuel cracking in the oil industry. And second, phenomena related to the magnetic shape memory effect in ferromagnetic shape memory alloys, a class of active materials used for components in actuators or dampers. The final outlook points out what parts of the the presented modeling approach might be subject to further analysis and investigations, and what new problems can be treated with the newly developed and implemented solution methods.

The appendix contains additional information and can be useful to gain a deeper understanding of some of the solution methods and implementations presented in this work. It contains discussions about the interpretation of orientations in the context of the software framework and how unit quaternions can serve as an alternative implementation for rotation matrices. The representation of Hooke's law of linear elasticity in a six-dimensional vector space will be presented, as well as a simplified compact notation for a numerical update scheme presented to compute the mechanical equilibrium condition.

Part I

Basics

2 Mathematics and notations

This chapter introduces the basic notations that are used throughout this text. It follows the structure of classic text books, but the author took the freedom to not always give definitions in the most general form and to omit rigorous proofs where it seemed appropriate.[1] The usual infix notation is used for operations and 'multiplications dots' are omitted wherever it increases the readability of the text[2]. Definitions and notations are limited to the extend required for this work. The reader is assumed to have a basic intuitive knowledge of mathematics, set theory and integration theory. The key idea of this chapter is to start from few basic principles and ideas, and to show how these can be used to classify 'real physical structures'. Examples are the classification of crystals and their symmetries, and the determination of the number of twin variants in martensitic materials and their categorization, what is done by finding the solutions to an algebraic equation whose solutions are related to the allowed directions of planes that separate two twins and the shear movement relative to this plane. The classifications are done by transferring the physical (and experimentally observable) properties and assumptions into an abstract mathematical framework that provides the appropriate methods for the classification process. The author tried to motivate the definitions in this chapter by stating where they will be used in the process of modeling the material behavior treated in this text.

---

[1] For instance, in an abstract group the identity element is unique and the left-inverse elements coincide with the right-inverse elements. These facts are not proven but included in the definitions.

[2] e.g. when multiplications are applied or maps are concatenated

2.1 Algebraic structures

This first section defines the algebraic structures that play a role in this text. This includes the concept of abstract groups that appears in the formalization of crystal structures and crystal symmetries, and Lie-groups and their associated Lie-algebras as special structures appearing in geometric integration. But also linear algebra, vector spaces and some properties of Fourier transforms will be stated and discussed.

Maybe groups are the most simple mathematical structures in which a 'multiplication operation' can be intuitively defined: There exists an element that 'does nothing' (the identity), every operation can be reversed (every element has an inverse) and the composition of two elements of the group stays in the group (a group is closed under the multiplication operation). This chapter follows the basic definitions and properties of finite groups in the context of the classification of crystals that is given in the text book of Bradley and Cracknell [13]. In the book of Kurzweil and Stellmacher [14] finite groups are classified. The first definition states what is understood by a group. The standard notation for the cardinality of a set is applied there: For any set M, by $|M|$ its *cardinality*, that is the number of elements in M, is denoted. So, if M is the *empty set*, $|\emptyset| = 0$. If $|M| \in \mathbb{N}$, the set M is a *finite set*, and an *infinite set* if $|M| = \infty$ (without differentiating if the set has countable or uncountable many elements).

Definition 2.1 (Groups) A *group* $(G, \cdot)$ is a set G together with an inner map (called *multiplication*) $\cdot : G \times G \to G$, such that

$$\forall g_1, g_2, g_3 \in G : g_1 \cdot (g_2 \cdot g_3) = (g_1 \cdot g_2) \cdot g_3 \qquad \text{(associativity)}$$

$$\exists! e \in G \; \forall g \in G : e \cdot g = g = g \cdot e$$
$$\text{(existence of an \textit{identity element})}$$

$$\forall g \in G \; \exists! g^{-1} \in G : g^{-1} \cdot g = e = g \cdot g^{-1}$$
$$\text{(existence of \textit{inverse elements})}$$

If all elements commutate, the group is a *commutative* or *abelian group*[3]. G is a *finite group*, if $|G| < \infty$, and an *infinite group* otherwise. The

---

[3]Named after the Norwegian mathematician Niels Henrik Abel ($\star$August 5th 1802 - $\dagger$April 6th 1829).

number of elements in G is the *order of the group*. A subset $G' \subseteq G$ is a *subgroup* of G, if G' is itself a group with respect to the multiplication '·', symbolized by $G' \leq G$. □

If G is a finite group with n elements, it can be completely defined by a *multiplication table*, i.e. a square scheme where the elements $e, g_1, \ldots, g_{n-1}$ are the labels for the rows and columns, and the products $g_i g_j$ of the elements are listed. Further, if a group is abelian, one usually denotes the group multiplication by the symbol '+' rather than the multiplication dot. Using the multiplication map of a group, the group can be subdivided into substructures. The properties related to these substructures simplify analysis and classification of abstract groups.

Definition 2.2 (COSETS AND FACTOR GROUPS) Let G be a group, $H \leq G$ and $g \in G$. The set $gH = \{gh | h \in H\}$ is called a *left coset* of H and $Hg = \{hg | h \in H\}$ is called a *right coset* of H. The number of different left and right cosets coincides and is called the *index* of H in G, written $[G : H]$. The set of all left cosets of H in G is denoted by G/H. If all left and right cosets of H in G coincide (i.e. if for all $g \in G$: $gH = Hg$), the subgroup H is a *normal subgroup* (or an *invariant subgroup*) of G, written $H \trianglelefteq G$. In this case, G/H itself can be equipped with a group structure via the multiplication

$$\forall gH, g'H \in G/H : (gH)(g'H) := (gg')H.$$

G/H is a *quotient group* or *factor group*.
For $g \in G$ the *conjugation with g* is defined by

$$\varphi_g : G \to G, g' \mapsto gg'g^{-1}.$$

So, H being a normal subgroup is equivalent to H being closed under conjugation with elements of G, i.e. $\varphi_g(H) = H$ for all $g \in G$. □

The following theorem is one of the basic theorems of group theory. In finite groups, it allows to count the number of cosets generated by a subgroup (cp. [14]).

Theorem 2.1 (THEOREM OF LAGRANGE) Let G be a finite group and $H \leq G$. Then for the index of H in G the relation

$$[G : H] = \frac{|G|}{|H|} \in \mathbb{N}$$

holds. Especially, the number of elements in H is a divisor of the number of elements in G. PROOF. For all $g \in G$ the left multiplication with g

$$H \to gH, h \mapsto gh$$

is a bijection, so $|H| = |gH|$. The cosets of H in G are a partition of G, and hence two cosets are either equal or disjoint:

$$G = \bigcup_{g \in G} gH \qquad \text{and} \qquad \forall g, g' \in G : gH \cap g'H \neq \emptyset \to gH = g'H.$$

So $|G| = |\bigcup_{g \in G} gH| = n \cdot |H|$, where n is the number of different left cosets of H in G. $\qquad\qquad\qquad\qquad\qquad\qquad\qquad\qquad\qquad\qquad\qquad\quad \square$

The concept of a group acting on a set substantiates the idea that a group 'does something', rather then being just an abstract set equipped with an inner map. For example, the set of transformations of a given structure forms a group when the composition of maps is taken as the group multiplication. One usually tends to think of concrete transformations, e.g. of the symmetry operations of a regular polygon or a crystal structure, or the infinite number of rotations around a fixed axis that map a sphere back onto itself. These ideas are summed up in the notion of *group actions.*

Definition 2.3 (GROUP ACTIONS) Let G be a group with identity e and let M be a set. A map $\Lambda : G \times M \to M$ satisfying

$$\forall x \in M : \Lambda(e, x) = x \quad \text{(identity acts trivial)}$$

$$\forall g_1, g_2 \in G \; \forall x \in M : \Lambda(g_1, \Lambda(g_2, x)) = \Lambda(g_1 g_2, x) \qquad \text{(associativity)}$$

is an *action of G on M.*
The action is a *transitive action,* if

$$\forall x, y \in M \; \exists g \in G : \Lambda(g, x) = y,$$

and the action is a *free action*, if

$$\forall g \in G : (\forall x \in M : \Lambda(g,x) = x \rightarrow g = e),$$

i.e. if no point of M is fixed by the action of any element different from the groups identity element e. $\square$

The definition shows that the action of a group on a set respects the group multiplication. If a group acts on a set, then there might be elements of the set that stay fixed under the operation of certain elements of the group that are not the identity (i.e. a group action can be non-free). Taking as an example the group of rotations acting on a Euclidean space, then for an arbitrary rotation each point on the axis of rotation stays fixed, while all other elements of the space are moved. The next definition makes these ideas more concrete.

Definition 2.4 (STABILIZERS, ORBITS AND CONJUGATION CLASSES)
Let G be a group that acts on a set M via a group action Λ. The *stabilizer* (or *fix point group*) of $m \in M$ is the set consisting of the elements of G that leave m fixed:

$$G_m = \{g \in G | \Lambda(g,m) = m\}.$$

The *orbit* of m is the set of points 'reachable from m' by applying elements of G:

$$O_m = \{\Lambda(g,m) | g \in G\}.$$

G_m is a subgroup of G, and O_m is a subset of M.
If $M = G$ and Λ is the action by conjugation, then for $g \in G$ the orbit

$$O_g = \{\Lambda(h,g) | h \in G\} = \{\varphi_h(g) | h \in G\} = \{hgh^{-1} | h \in G\}$$

is the *conjugacy class* of g. $\square$

So, a group G acts free on a set M, if all stabilizers of elements in M are trivial (i.e. equal to $\{e\}$). The number of elements in the orbit of a point is a divisor of the number of elements in the group. This is a consequence of the following theorem in combination with Thm. 2.1.

Theorem 2.2 (ORBIT-STABILIZER THEOREM) Let G be a finite group that acts via Λ on a set M, and let $m \in M$. Then

$$|O_m| = [G : G_m].$$

PROOF. The short proof follows [14]. Let $g, h \in G$. Then

$$\Lambda(g, m) = \Lambda(h, m) \leftrightarrow \Lambda(h^{-1}g, m) = m \leftrightarrow h^{-1}g \in G_m \leftrightarrow g \in hG_m.$$

This shows that the number of different elements 'reachable from m' with elements of G is the same as the number of different cosets of G_m in G, i.e. $|O_m| = [G : G_m]$. $\qquad\square$

With Thm. 2.1 easily follows as a corollary the proposition stated above.

Corollary 2.1 (ORBIT LENGTH DIVIDES THE GROUP ORDER) Let G be a finite group that acts via an action Λ on a set M, and let $m \in M$. Then

$$|O_m| = [G : G_m] = \frac{|G|}{|G_m|} \in \mathbb{N},$$

and therefore $|O_m|$ is a divisor of $|G|$. $\qquad\square$

Groups can be combined to gain new groups. For example, the *Euclidean group*, that is the group of all distance and angle preserving maps of a real space, can be recognized as the semi-direct product of the group of rotations and reflections with the group of translations.

Definition 2.5 (DIRECT AND SEMI-DIRECT PRODUCTS OF GROUPS) Let G and H be groups. The *direct product $G \times H$* of G and H is defined as

$$(g_1, h_1)(g_2, h_2) \mapsto (g_1g_2, h_1h_2),$$

and $G \times H$ is again a group.

If $(H, +)$ is abelian and if the group G acts linearly on H via a group action Λ, then the *semi-direct product, $G \ltimes H$,* is defined by

$$(g_1, h_1)(g_2, h_2) \mapsto (g_1 g_2, \Lambda(g_1, h_2) + h_1),$$

and $G \ltimes H$ is again a group. $\qquad\qquad\square$

In all fields of mathematics, the analysis of *structure preserving maps* (that are maps from one algebraic structure to another that do not change the defining properties of the structure) is an important tool. Properties can be derived from other, sometimes well-known, structures, or insights be gained within certain substructures already contained in the structure itself. The following definition refers only to groups for simplicity.[4]

Definition 2.6 (GROUP HOMOMORPHISMS AND KERNELS) Let G and H be groups, and let $\varphi : G \to H$ be a map. φ is called a *homomorphism*, if

$$\forall g, g' \in G : \varphi(gg') = \varphi(g)\varphi(g').$$

If φ is bijective, then φ is an *isomorphism*, and G and H are *isomorphic*. The *image* of φ is the set of images of elements of G in H, i.e. $\varphi(G) := \{\varphi(g)|g \in G\}$, and the *pre-image* $\varphi^{-1}(h)$ of $h \in H$ is the set of elements of G mapped to h, i.e. $\varphi^{-1}(h) = \{g \in G|\varphi(g) = h\}$. The *kernel* of φ is the set of elements mapped onto the identity e_H of H by φ, i.e. $\ker\varphi := \{g \in G|\varphi(g) = e_h\}$. The set $\ker\varphi$ is a normal subgroup of G.
$\square$

The notations of image and pre-image apply to all functions and are not restricted to (group) homomorphisms, while the definition of the kernel needs an underlying (group) structure as it refers to a dedicated element (the identity element).

Algebraic structures that extend the possibilities to do the 'usual calculations' by combining a multiplication and an addition are *fields*. They

---

[4]The transfer of the next definition to other algebraic structures than groups is straight forward.

connect two group structures assigned to a set via the laws of distributivity. If the multiplication is non-abelian, the structure is a skew field. Skew fields play a crucial role in the representation of rotations using so called unit quaternions, as rotations in general do not commute (see Appendix A.4 for more detailed explanations).

Definition 2.7 (SKEW FIELDS AND FIELDS) Let $\mathbb{K}$ be a set and $0 \in \mathbb{K}$. Let $\cdot : \mathbb{K} \to \mathbb{K}$ and $+ : \mathbb{K} \to \mathbb{K}$ be two inner maps (called multiplication and addition) such that $(\mathbb{K} \setminus \{0\}, \cdot)$ is a group and $(\mathbb{K}, +)$ is an abelian group with identity 0. If

$$\forall k_1, k_2, k_3 \in \mathbb{K} : k_1 \cdot (k_2 + k_3) = (k_1 \cdot k_2) + (k_1 \cdot k_3), \quad \text{(distributivity)}$$

$(\mathbb{K}, \cdot, +)$ is a *skew field*. If $(\mathbb{K}, \cdot)$ is abelian, then $(\mathbb{K}, \cdot, +)$ is a *field*. $\qquad\square$

Vector spaces are often intuitively used algebraic structures. A proper definition needs an underlying field whose elements are called *scalars*. Finite-dimensional vector spaces are uniquely determined.[5] One of the most important vector spaces is the real three-space (i.e. the set of all three-tuples of real numbers) as it is usually used to represent the ambient space. In mechanics, other vector spaces become important (e.g. the space of real six-tuples in the matrix representation of Hooke's law of linear elasticity, see Appendix B), so a more general definition will be given here.

Definition 2.8 (VECTOR SPACES) Let $\mathbb{K}$ be a field and V be an abelian group. V is a $\mathbb{K}$-*vector space*, if the *scalar multiplication*

$$\cdot : \mathbb{K} \times V \to V$$

satisfies

(i) $\forall a, b \in \mathbb{K} \; \forall v \in V : a \cdot (b \cdot v) = (ab) \cdot v$

(ii) $\forall a \in \mathbb{K} \; \forall v, w \in V : a \cdot (v + w) = (a \cdot v) + (a \cdot w)$

(iii) $\forall a, b \in \mathbb{K} \; \forall v \in V : (a + b) \cdot v = (a \cdot v) + (b \cdot v)$

---

[5]Up to, as usual in mathematics, isomorphisms.

Let V be a $\mathbb{K}$-vector space, $n \in \mathbb{N}$ and $\{v_1, \ldots, v_n\} \subseteq V$. The v_i are *linearly independent*, if for all $(a_1, \ldots, a_n) \in \mathbb{K}^n$

$$\sum_{i=1}^{n} a_i v_i = \mathbf{0} := (0, \ldots, 0)^T \rightarrow a_1 = \cdots = a_n = 0.$$

Otherwise, the v_i are *linearly dependent*. A maximal set[6] of linearly independent vectors $B \subseteq V$ is a *basis* of V, and V is a *finite-dimensional vector space* of *dimension* n if $|B| = n$. All bases have the same cardinality, and each vector $v \in V$ can be uniquely expressed as a linear combination of the elements of a basis. $\qquad\square$

The above definition of a basis can be extended to infinite dimensional vector spaces, and some sets of functions can be equipped with a vector space structure of infinite dimension. For each vector space, a basis can be found. This is stated by the next theorem, which will not be proven here, because in the case of infinite dimensional vector spaces the *axiom of choice*[7] is required (in form of the 'Lemma of Zorn'). For a proof the reader is referred to the textbook of Bosch [16].

Theorem 2.3 (EXISTENCE OF BASES) Let V be a vector space. Then V has a basis. $\qquad\square$

In the same way as defined for groups in Def. 2.6, structure preserving maps are defined for other algebraic structures such as *Lie-group homomorphisms, vector space homomorphisms, field homomorphisms* etc. The kernel of every homomorphism is the set of elements mapped onto the identity. Kernel and image of a homomorphism always respect the algebraic structure. Because vector spaces play a special role, some details are given explicitly.

---

[6]maximal with respect to set inclusion

[7]The *axiom of choice* states that for each set of non-empty sets there exists a function that chooses one element from each of these sets. This idea is easy to describe and often intuitively assumed to be true, but has important non-trivial consequences. See e.g. the book of Deiser [15] for detailed explanations.

Definition 2.9 (LINEAR MAPS AND LINEAR OPERATORS ON VECTOR SPACES) Let V, W be $\mathbb{R}$-vector spaces of dimension n and m, and let B_V and B_W be fixed bases of V and W. A vector space homomorphism $A : V \to W$ is called a *linear map*. If A is bijective, then $n = m$ and A describes the change of coordinates, as $A(B_V)$ is again a basis. The set of all linear maps from V to W is denoted by $Lin(V, W)$, and is itself a vector space via

$$\forall T, T' \in Lin(V, W)\ \forall v \in V : (T + T')v := T(v) + T'(v)$$

and

$$\forall T \in Lin(V, W)\ \forall r \in \mathbb{K}\ \forall v \in V : (rT)(v) := r(T(v)).$$

If $V = W$, the elements $T \in Lin(V, W)$ are *linear operators*, as they operate on the underlying vector space $\mathbb{R}^n$. If $\ker T = \{\mathbf{0}\}$, the operator T is a *non-singular operator*, and a *singular operator* otherwise. $\square$

If V is a finite dimensional vector space of dimension n with basis $B_V = \{b_1^V, \ldots, b_n^V\}$, V becomes canonically isomorphic to the underlying tuple space $\mathbb{K}^n$ by identifying each vector with the coefficients as a linear combination of the (ordered) basis B_V: So the spaces $\mathbb{R}^n$ are the only examples of n-dimensional real vector spaces. The reader should keep in mind that fixing a basis is crucial when the entries of a vector are interpreted. A linear map between V and a finite dimensional vector space W with basis B_W can thus be represented in a rectangular matrix scheme $A = (a_{ij})_{i=1\ldots n}^{j=1\ldots m}$, where the i-th row of A contains the coefficients of $A(b_i^V)$ as a linear combination of the base vectors b_j^W of W. The matrix A^T is the matrix gained from A by exchanging rows and columns and called the *transpose* of A. Sometimes it will be convenient to interpret *column vectors* $v \in \mathbb{R}^n$ as $n \times 1$-matrices, so that the vector v^T is a *row vector*. If $n = m$, two special matrices are defined: The matrix $\mathbf{I} = (e_{ij})$ with $e_{ij} = 1$ if $i = j$, and $e_{ij} = 0$ if $i \neq j$ is the $n \times n$ *identity matrix*, and the matrix $\mathbf{0} = (o_{ij})$ with $o_{ij} = 0$ for all $i, j \leq n$ is the $n \times n$ *zero matrix*.

Groups are until now an abstract concept. Since material properties and crystals shall be characterized using this concept, now the representation of a group will be defined. A representation of an abstract groups allows to 'fill' the abstract concept with a less abstract 'view' without losing the group properties. The definition follows [13].

Definition 2.10 (REPRESENTATIONS OF GROUPS) Let G be a group and $\mathbf{T}$ be a group of non-singular linear operators that act on a finite dimensional real (or complex) vector space V (that is $\mathbf{T} \leq Lin(V, V)$). A homomorphism

$$\gamma : G \to \mathbf{T}, g \mapsto T_g$$

is called a *representation* of G. If γ is an isomorphism, the representation is a *faithful representation*. Let $B = \{b_1, \dots, b_n\}$ be a basis of V. Then for all $g \in G$ matrices $T_B(g)$ can be defined by the equations

$$T_g(b_i) = \sum_{j=1}^{n} (T_B(g))_{ij} b_j \qquad i = 1, \dots, n.$$

$T_B(g)$ is the *matrix representation* of $g \in G$ with respect to the basis B in the representation given by γ. $\qquad\qquad\square$

In general, representations are not unique. In this work, abstract groups will be identified with concrete well-known and commonly used matrix representations, what is in agreement with many text books. An example is the identification of the set of all rotations of a finite-dimensional real vector space with the set of matrices having unit determinant and for which the inverse and the transpose coincide.

The following examples serve several purposes: They illustrate the abstract structures that are introduced in this section, and give the commonly used notations for these structures. Some representations of often occurring abstract groups will be defined. Later on in this text, abstract groups will be identified with concrete matrix representations.

Example 2.1 (ALGEBRAIC STRUCTURES)

1. The set of real numbers $(\mathbb{R}, +, \cdot)$ with the usual addition '$\cdot$' and multiplication '$+$' is a field.

2. The set of complex numbers $\mathbb{C} = \{a + ib \,|\, a, b \in \mathbb{R}\}$ is a field, where the 'imaginary unit' i is a solution of the equation $x^2 = -1$. This field is isomorphic to the set of 2×2 matrices

$$V_{\mathbb{C}} = \left\{ \begin{pmatrix} a & -b \\ b & a \end{pmatrix} \,|\, a, b \in \mathbb{R} \right\}$$

via the isomorphism

$$\varphi : \mathbb{C} \to V_{\mathbb{C}}, a + ib \mapsto \begin{pmatrix} a & -b \\ b & a \end{pmatrix}.$$

So, $V_{\mathbb{C}}$ can be equipped with the structure of a field, and the relations

$$\varphi(0) = \mathbf{0}, \quad \varphi(1) = \mathbf{I}, \quad \varphi(i) = \begin{pmatrix} 0 & -1 \\ 1 & 0 \end{pmatrix} \quad \text{and} \quad (\varphi(i))^2 = -\mathbf{I}$$

hold.

3. Let $\mathbb{K}$ be a field and $n \in \mathbb{N}$. The set of all ordered n-tuples $\mathbb{K}^n$ is a vector space with component-wise scalar multiplication. The vector spaces $\mathbb{K}^n$ are the only examples of finite vector spaces. The three-dimensional vector space over the reals $\mathbb{R}^3$ has the *standard basis* $\mathcal{B} = \{(1,0,0), (0,1,0), (0,0,1)\}$. This definition directly transfers to $\mathbb{R}^n$ with $n \neq 3$.

   The set $\mathbb{K}^{n \times n}$ of all $n \times n$ matrices $A = (a_{ij})_{i,j=1}^{n}$ is a vector space with component-wise scalar multiplication $kA := (ka_{ij})$ for all $k \in \mathbb{K}$.

   With $\mathbb{Z}^{3 \times 3}$, the set of 3×3 integer matrices is denoted. The reader should remind that $\mathbb{Z}$ is not a field.

4. This text frequently refers to the following matrix groups that are representations of (informally) defined abstract groups:

   - $\mathrm{GL}(n, \mathbb{R})$: The *general linear group* is the group of invertible $n \times n$-matrices over the reals.

   - $\mathrm{SL}(n, \mathbb{R})$: The *special linear group* is the group of $n \times n$-matrices with unit determinant.

   - $\mathrm{O}(n, \mathbb{R})$: The *orthogonal group* is the group of $n \times n$-matrices $R \in \mathrm{GL}(n, \mathbb{R})$ for which the relation $R^T = R^{-1}$ is valid. This represents the set of all angle and distance preserving linear maps of $\mathbb{R}^n$, i.e. rotations and reflections.

- SO$(n, \mathbb{R})$: The *special orthogonal group* is the group of $n \times n$-matrices $R \in$ O$(n, \mathbb{R})$ with unit determinant. This represents all rotations of $\mathbb{R}^n$.
- E$(n) =$ O$(n, \mathbb{R}) \ltimes \mathbb{R}^n$ is the *Euclidean group* of rigid body motions.
- SE$(n) =$ O$(n, \mathbb{R}) \ltimes \mathbb{R}^n$ is the *special Euclidean group* of orientation preserving rigid body motions.
- symm$(\mathbb{R}^{n \times n}) = \{A \in \mathbb{R}^{n \times n} | A = A^T\}$ is the set of real *symmetric $n \times n$-matrices.*
- skew$(\mathbb{R}^{n \times n}) = \{A \in \mathbb{R}^{n \times n} | A = -A^T\}$ is the set of real *skew-symmetric $n \times n$-matrices.*

$$\square$$

The following relations hold:

$$\text{SO}(n, \mathbb{R}) \trianglelefteq \text{O}(n, \mathbb{R}) \leq \text{SL}(n, \mathbb{R}) \leq \text{GL}(n, \mathbb{R}) \leq Lin(\mathbb{R}^n, \mathbb{R}^n).$$

For $n = 3$ these groups act on the space of experience, the real three-space, via matrix-vector multiplication.

The above introduced notations can be generalized to fields different from $\mathbb{R}$. As this is not needed in this work and mostly the field $\mathbb{R}$ will be considered, the field dependency will often be omitted, leading to the abbreviating notation GL(n), O(n), SO(n) etc.

This first section is concluded by the definition of a special action of GL$(n, \mathbb{R})$ on $\mathbb{R}^n$ that describes the change of the basis of a vector space.

Definition 2.11 (SIMILARITY TRANSFORMATIONS) Let $n \in \mathbb{N}$ and $R \in$ GL$(n, \mathbb{R})$. GL$(n, \mathbb{R})$ acts on $Lin(\mathbb{R}^n, \mathbb{R}^n)$ via matrix multiplication. The action of conjugation with R, i.e.

$$\varphi_R : Lin(\mathbb{R}^n, \mathbb{R}^n) \to Lin(\mathbb{R}^n, \mathbb{R}^n), A \mapsto RAR^{-1} = A'.$$

is called *similarity transformation.* $\qquad\square$

The matrix A' in the above definition describes the effect of the transformation $A \in Lin(\mathbb{R}^n, \mathbb{R}^n)$, after a change of bases, determined by $R \in$ GL$(n, \mathbb{R})$, has been applied.

2.2 Properties of vector spaces

The structures defined in the previous section will now be used to characterize basic properties of vector spaces and point lattices. These properties are commonly used to describe various physical settings, as they appear naturally in the perception of an observer of such a system.

For this section, let $n \in \mathbb{N}$ be fixed. The finite n-dimensional vector space over the reals, $\mathbb{R}^n$, is considered here. Basic knowledge of the ordering of the field of the reals is assumed, as well as on the integrability of functions.

Definition 2.12 (ABSOLUTE VALUE OF REAL NUMBERS) Let $r \in \mathbb{R}$. The *absolute value* of r is defined as

$$|r| = \begin{cases} r & \text{if } r \geq 0 \\ -r & \text{if } r < 0 \end{cases}.$$

$\square$

The absolute value of a real induces the measure of the distance between two real values as the absolute value of their difference. A generalization of this concept are norms which induce a measure of the distance between vectors.

Definition 2.13 (NORMS, p-NORMS AND SPHERES) A *norm* on $V = \mathbb{R}^n$ is a map

$$|| \cdot || : V \to \mathbb{R}_{\geq 0}$$

satisfying

(i) $\forall v \in V : ||v|| = 0 \to v = \mathbf{0}$

(ii) $\forall v \in V \; \forall k \in \mathbb{R} : ||kv|| = |k| \, ||v||$

(iii) $\forall v, w \in V : ||v + w|| \leq ||v|| + ||w||$

Let $p \in \mathbb{R}$. Then

$$\forall v = (v_1, \ldots, v_n)^T \in R^n : ||v||_p := \left(\sum_{i=1}^{n} |v_i|^p\right)^{\frac{1}{p}}$$

is a norm, the so called *p-norm* on $\mathbb{R}^n$. The 2-norm of $v \in \mathbb{R}^n$ is the *Euclidean norm* or the *length* of v, written $|v| := ||v||_2$.
The set $\mathbb{S}^{n-1} := \{x \in \mathbb{R}^n \mid |x| = 1\}$ is the *n-dimensional unit sphere* in $\mathbb{R}^n$. $\qquad\qquad\square$

Some commonly used notations on real vector spaces will be introduced in the next definition.

Definition 2.14 (SCALAR PRODUCT, CROSS PRODUCT AND ORTHOGONALITY) Let $v, w \in \mathbb{R}^n$. The *scalar product* (or *dot product*) of v, w is defined as

$$v \cdot w = \sum_{i=1}^{n} v_i w_i.$$

If $v \cdot w = 0$, then v, w are *orthogonal*.
If $n = 3$ the *cross product* of v and w is defined as

$$v \times w = (v_2 w_3 - v_3 w_2, v_3 w_1 - v_1 w_3, v_1 w_2 - v_2 w_1)^T.$$

The following geometric relations to trigonometric functions are valid:

$$v \cdot w = |v||w| \cos \angle(v, w) \qquad \text{and} \qquad v \times w = |v||w| \sin \angle(v, w) n,$$

where $\angle(v, w)$ denotes the angle between v and w, and $n \in \mathbb{S}^2$ is the vector orthogonal to the plane in which v and w lie, i.e. $v \cdot n = 0 = w \cdot n$. $\qquad\square$

The scalar product of $v, w \in \mathbb{R}^n$ can equivalently be written as the matrix-matrix product $v^T w$.

Definition 2.15 (ORTHONORMAL BASES, FRAMES OF REFERENCE AND RECIPROCAL BASES) Let $V = \mathbb{R}^n$ and let $B = \{b_1, \ldots, b_n\}$ be a basis of V. If

$$b_i \cdot b_j = \begin{cases} 0 & \text{if } i \neq j \\ 1 & \text{if } i = j \end{cases},$$

B is a *orthonormal basis.*

If the vectors of B are centered at the point $\mathbf{o} \in V$, then $C = (\mathbf{o}, b_1, \ldots, b_n)$ forms a *Cartesian coordinate system* or *frame of reference* of V, and every element of $v \in C$ can be written uniquely as

$$v = \sum_{i=1}^{n} r_i b_i + \mathbf{o} \text{ with } (r_1, \ldots, r_n)^T \in \mathbb{R}^n.$$

For each basis B the *reciprocal basis* $B^r \{b^1, \ldots, b^n\}$ to B is defined via

$$b_i \cdot b^j = \begin{cases} 0 & \text{if } i \neq j \\ 1 & \text{if } i = j \end{cases}.$$

$\square$

Every finite-dimensional real vector space can be equipped with an orthonormal basis, as the next proposition states.

Theorem 2.4 (EXISTENCE OF ORTHONORMAL BASES) Let $V = \mathbb{R}^n$. Then $B \subseteq V$ exists, such that B is an orthonormal basis of V.
PROOF. The orthonormalization scheme by Gram and Schmidt (see eg. [16]) is a constructive method to find a basis of V with the desired properties. $\square$

Convention If not stated otherwise, all appearing real vector spaces are assumed to be equipped with an orthonormal basis.

An important set of linear maps is generated by so called dyadic products of two vectors. These products become fundamental in the description of shear deformations in twinned microstructures and more generally in the coordinate free description using tensors.

Definition 2.16 (DYADIC PRODUCTS) Let $s, n \in \mathbb{R}^n$. The *dyadic product* of s and n is the linear map

$$s \otimes n : \mathbb{R}^n \to \mathbb{R}^n, v \mapsto (n \cdot v)s.$$

It is easy to see from the definition that the image $(s \otimes n)(\mathbb{R}^n) = \{\alpha s | \alpha \in \mathbb{R}\}$ is one-dimensional, and that the kernel $\ker(s \otimes n)$ is $(n-1)$-dimensional. $(s \otimes n)$ can equivalently be written as sn^T. $\qquad\qquad\square$

When a basis is fixed, the set of all matrices $A = (a_{ij})$ is itself a vector space (or is, more precisely, isomorphic to $Lin(\mathbb{R}^n, \mathbb{R}^n)$). If orthonormal bases are fixed, the concept of a tensor is defined by its transformation under *proper rotations*[8]. Tensors are very often used in physics to represent *anisotropic material properties* (i.e. properties that might differ in different crystallographic directions, see [17]), such as the mechanical stress or strain and elastic stiffness in the theory of elasticity, but also in magnetism, piezoelectrics etc. The following definition of a tensor is based on the book of Neumann and Schade [18]. It uses an abbreviating notation that is often applied in physics and mechanics.

Convention It is common in the literature to suppress the summation sign $\sum$ and implicitly sum over repeatedly occurring indexes. This so called *Einstein summation convention* makes formulae more compact. As it sometimes may cause confusion, it has to be used carefully. In this work, the convention is generally avoided with some very few exceptions.

Definition 2.17 (TENSORS) Let V be a real n-dimensional vector space. Let $N, n \in \mathbb{N}$ be natural numbers and $R = (r_{ij}) \in \mathrm{GL}(n)$. The numbers $T_{i_{k_1}, \ldots, i_{k_N}} \in \mathbb{R}$ with $(i_{k_1}, \ldots, i_{k_N}) \in \{1, \ldots, n\}^N$ define a *tensor* T of

---

[8]Proper rotations are orientation, distance and angle preserving maps, i.e. elements of $\mathrm{SO}(n)$.

rank N and order n, if (applying Einstein's summation convention on all n-tuples $(j_{k_1}, \ldots, j_{k_N}) \in \{1, \ldots, n\}^N$)

$$T'_{i_{k_1}, \ldots, i_{k_N}} = r_{i_{k_1} j_{k_1}} \ldots r_{i_{k_N} j_{k_N}} T_{j_{k_1}, \ldots, j_{k_N}} \qquad \text{(transformation of tensors)}$$

is valid for the coordinate transformation according to R. The $T'_{i_{k_1}, \ldots, i_{k_N}}$ are the entries of a tensor T' in the new coordinate system with n^N entries.

$\square$

So, every tensor is represented by a matrix, but not every matrix represents a tensor. As an $n \times m$ real matrix can be thought of as an element of $\mathbb{R}^{n \cdot m}$, the definition of norms and scalar products can be naturally extended to tensors: Let $A, B \in \mathbb{R}^{n \times m}$. Then

$$|A| = ||A||_2 = \sqrt{\sum_{i=1}^{n} \sum_{j=1}^{m} a_{ij}^2}$$

and

$$A \cdot B = \sum_{i=1}^{n} \sum_{j=1}^{m} a_{ij} b_{ij}.$$

Remark The tensors appearing in this work mostly are of order $n = 3$. For their rank N, usually $N \in \{0, 1, 2, 3, 4\}$, where $N = 0$ refers to scalars and $N = 1$ to vectors. For $N = 2$ matrix schemes can be used to write the tensors, for $N > 2$ it becomes more difficult to visualize the tensors. Examples for second rank tensors are the mechanical stresses and strains (with nine entries each), and the elastic stiffness and compliance tensors (with 81 entries) are examples of fourth order tensors.

Remark For $N = 2$ the transformation law for tensors in Def. 2.17 is the similarity transformation from Def. 2.11: Let $A = (a_{ij}) \in \mathrm{GL}(n)$ and $R = (r_{ij}) \in \mathrm{SO}(n)$. Then

$$a'_{ij} = r_{ik} r_{jl} a_{kl} = r_{ik} a_{kl} r_{jl} = r_{ik} a_{kl} r^T_{lj},$$

where r_{lj}^T is the lj-th entry of R^T. Therefore the above formula gives the components of $A' = RAR^T = RAR^{-1}$ (because $R \in \mathrm{SO}(3)$).

In the case $N = 4$, the general transformation law explicitly reads

$$T'_{ijkl} = r_{ip}r_{jq}r_{kr}r_{ls}T_{pqrs}.$$

Here, the Einstein summation convention is applied.

The following definition and decomposition theorem are essential to identify the parts of transformations that are relevant to describe the action of linear operators. It is needed in mechanics, when observer independence is discussed (see [10]).

Definition 2.18 (POSITIVE DEFINITENESS) Let $n \in \mathbb{N}$ and $M \in \mathbb{R}^{n \times n}$. M is *positive definite* if

$$\forall v \in \mathbb{R}^n \setminus \{\mathbf{0}\} : v^T(Mv) > 0.$$

$\square$

Theorem 2.5 (POLAR DECOMPOSITION) Let $F \in \mathbb{R}^{3 \times 3}$ with $\det(F) > 0$. Then there exist a rotation $Q \in \mathrm{SO}(3)$ and a positive-definite $U \in \mathrm{symm}(\mathbb{R}^{3 \times 3})$ such that

$$F = QU.$$

The matrices Q and U are uniquely determined.

PROOF. Following [10], the matrices U and Q will be constructed:

Let $C = F^T F$. Then $C \in \mathrm{symm}(\mathbb{R}^{3 \times 3})$ and, because of $\det(F) > 0$, positive definite. Therefore, C has the three different positive eigenvalues $\gamma_1, \gamma_2, \gamma_3 \in \mathbb{R}$. Let $u_1, u_2, u_3 \in \mathbb{R}^3$ be three corresponding and mutually perpendicular eigenvectors and set $\mu_i = \sqrt{\gamma_i} > 0$. Define the matrix

$$U = \sum_{i=1}^{3} \mu_i(u_i \otimes u_i).$$

As U has the same eigenvalues and eigenvectors as C, U is invertible. With $Q = FU^{-1}$ follows $F = QU$. The uniqueness of Q and U can be easily verified. $\hfill\square$

Derivatives of functions on vector spaces are a way to quantify how these functions locally change. The differential 'nabla operator' (or 'del operator') ∇ is an abbreviation used in mathematics and physics for different differential operations such as gradient, divergence, curl and related operations.

Definition 2.19 (THE NABLA OPERATOR) Let $\{e_i | i = 1, \ldots, n\}$ be the standard basis of $\mathbb{R}^n$. The *nabla operator* or *del operator* ∇ in $\mathbb{R}^n$ is defined as

$$\nabla = \sum_{i=1}^{n} \frac{\partial}{\partial x_i} e_i = \left(\frac{\partial}{\partial x_1}, \ldots, \frac{\partial}{\partial x_n} \right)^T.$$

Let $f : \mathbb{R}^n \to \mathbb{R}^m$ be a scalar-valued function. Then the *gradient* of f is

$$\nabla f = \begin{pmatrix} \frac{\partial}{\partial x_1} f_1 & \cdots & \frac{\partial}{\partial x_n} f_1 \\ \vdots & & \vdots \\ \frac{\partial}{\partial x_1} f_m & \cdots & \frac{\partial}{\partial x_n} f_m \end{pmatrix}.$$

For $m = 1$ this reads $\nabla f = \left(\frac{\partial}{\partial x_1} f \cdots \frac{\partial}{\partial x_n} f \right)$. For a vector field $F : \mathbb{R}^n \to \mathbb{R}^n$ the *divergence* of F is

$$\nabla \cdot F = \sum_{i=1}^{n} \frac{\partial}{\partial x_i} F_i.$$

The *Laplace operator* Δ is the divergence of the gradient, i.e.

$$\Delta F = \nabla \cdot \nabla F = \nabla^2 F = \sum_{i=1}^{n} \frac{\partial^2}{\partial x_i^2} F_i.$$

If $n = 3$, the *curl* of F is

$$\nabla \times F = \begin{pmatrix} \frac{\partial}{\partial x_2} F_3 - \frac{\partial}{\partial x_3} F_2 \\ \frac{\partial}{\partial x_3} F_1 - \frac{\partial}{\partial x_1} F_3 \\ \frac{\partial}{\partial x_1} F_2 - \frac{\partial}{\partial x_2} F_1 \end{pmatrix}.$$

$\square$

The next theorems state properties of functions defined on vector spaces. The first shows that any vector field can be decomposed into a curl-free part that consists of the negative gradient of a scalar potential, and a divergence-free part that consists of the rotation of a vector field (cp. the book of Jackson [19]).

Theorem 2.6 (HELMHOLTZ DECOMPOSITION THEOREM) Let $\Omega \subset \mathbb{R}^3$ be bounded and $F : \Omega \to \mathbb{R}^3$ be a vector field that is continuous on Ω and continuous and bounded on the surface boundary $\partial\Omega$. Then F can be completely decomposed into the sum of an *irrotational* field and a *solenoidal* field. I.e. there are a *scalar potential* $\psi : \mathbb{R}^3 \to \mathbb{R}$ and a vector field $A : \mathbb{R}^3 \to \mathbb{R}^3$, such that

$$F = -\nabla\psi + \nabla \times A.$$

PROOF. Let $\hat{\mathbf{n}}$ be the normal to $\partial\Omega$ pointing outwards. Define

$$\psi(x) = \frac{1}{4\pi} \int_\Omega \nabla \cdot F(x') \frac{x - x'}{|x - x'|} \, \mathrm{d}^3 x' - \\ \frac{1}{4\pi} \int_{\partial\Omega} \hat{\mathbf{n}}(x') \cdot F(x') \frac{x - x'}{|x - x'|} \, \mathrm{d}^2 x'$$

and

$$A(x) = \frac{1}{4\pi} \int_\Omega \nabla \times F(x') \frac{x - x'}{|x - x'|} \, \mathrm{d}^3 x' - \\ \frac{1}{4\pi} \int_{\partial\Omega} \hat{\mathbf{n}}(x') \times F(x') \frac{x - x'}{|x - x'|} \, \mathrm{d}^2 x'.$$

Then ψ and A meet the proposition. $\square$

Remark Take Ω and F as in the theorem above. If Ω is unbounded and F decays 'fast enough' for $|x| \to \infty$, then the surface terms in the above solutions vanish:

$$\psi(x) = \frac{1}{4\pi} \int_\Omega \nabla \cdot F(x') \frac{x - x'}{|x - x'|} \, \mathrm{d}^3 x'$$

and

$$A(x) = \frac{1}{4\pi} \int_\Omega \nabla \times F(x') \frac{x - x'}{|x - x'|} \, \mathrm{d}^3 x'.$$

Again, ψ and A meet the proposition.

The next theorem states that the action of a vector field inside a body can be described by the flux over the bodies boundaries (cp. [20]). It is also known as the *theorem of Gauss*.

Theorem 2.7 (DIVERGENCE THEOREM) Let $\Omega \subseteq \mathbb{R}^n$ be a region with boundary $\partial\Omega$. Let $F : \mathbb{R}^n \to \mathbb{R}^n$ be smooth in Ω. Then

$$\int_\Omega (\nabla \cdot F) \, \mathrm{d}\Omega = \int_{\partial\Omega} (F \cdot \hat{n}) \, \mathrm{d}\partial\Omega,$$

where $\hat{n}$ is the unit normal on $\partial\Omega$ pointing outwards. $\square$

The question how a given time- and space-dependent quantity changes when the mechanical body it is defined on deforms answers the following theorem. In the literature, there exist differently stated versions. A discussion about the different versions, including proofs, can be read in [21].

Theorem 2.8 (REYNOLDS TRANSPORT THEOREM) Let $\Omega \subset \mathbb{R}^3$ be a time-dependent material volume with surface $\partial\Omega$ of a mechanical body, and let $\psi : \Omega \times \mathbb{R}_{\geq 0} \to \mathbb{R}$ a 'time-dependent property'. Then

$$\frac{\mathrm{d}}{\mathrm{d}t} \int_\Omega \psi(x, t) \, \mathrm{d}\Omega = \int_\Omega \left(\frac{\partial \psi(x, t)}{\partial t} + \nabla \cdot (v\psi(x, t)) \right) \mathrm{d}\Omega$$

$$= \int_\Omega \frac{\partial \psi(x,t)}{\partial t}\, \mathrm{d}\Omega + \int_{\partial\Omega} \psi(x,t)(v \cdot \hat{n})\, \mathrm{d}\partial\Omega,$$

where v is the velocity of the flux over the surface boundary with normal $\hat{n}$. The second equality is a consequence of the first by means of the divergence theorem Thm. 2.7. $\qquad\square$

2.3 Euclidean motions and crystallography

This section defines crystallographic groups and point groups. The concept of discrete point lattices is used to define crystals, and methods used to denote directions in crystallographic structures will be presented here. Again, an $n \in \mathbb{N}$ is fixed for this section.

Rotations are the concept of transforming an n-dimensional vector space onto itself in a way that the distance between any two points in the space is kept unchanged, the chirality of the system is not affected and at least one point in space stays fixed. This makes rotations special kinds of isometries, that are angle and distance preserving transformations.

Definition 2.20 (ISOMETRIES IN $\mathbb{R}^n$) Let $R : \mathbb{R}^n \to \mathbb{R}^n$ be a map. If for all $x, x' \in \mathbb{R}^n$ the relation

$$x \cdot x' = R(x) \cdot R(x')$$

holds, R is called an *isometry*. If additionally R is a linear transformation, the isometry R is an *orthogonal transformation* and $|\det(R)| = 1$. If $\det(R) = 1$, R is a *rotation*, and if $\det(R) = -1$, R is a *reflection*. $\quad\square$

This definition is conform with the examples given at the end of the first section of this chapter, and as groups and their representations are identified, one can think of the set of orthogonal transformations as the matrix group $\mathrm{O}(n)$, and of the set of rotations as the group $\mathrm{SO}(n)$. The next theorem lists, without proofs, some simple properties of isometries that are needed later to characterize crystals and to construct a geometric integration method for the numerical integration of the evolution equation for the spontaneous magnetization.

Theorem 2.9 (CHARACTERIZATION OF ISOMETRIES AND ROTATIONS)
Let $R : \mathbb{R}^n \to \mathbb{R}^n$ be an isometry. Then:

   (i) R is bijective.

  (ii) $R \in \mathrm{E}(n) = \mathrm{O}(n) \ltimes \mathbb{R}^n$, that is R is the combination of a rotation or a reflection with a translation (i.e. R is a *Euclidean motion* or *rigid body motion*).

 (iii) If $R \in \mathrm{SO}(3)$, then the eigenvalues of R are 1, $\exp(i\varphi)$ and $\exp(-i\varphi)$ for a $\varphi \in [0, 2\pi[$. The eigenspace corresponding to the eigenvalue 1 is the *axis of rotation*, the parameter φ is the *angle of rotation* around this axis.

 (iv) The action of $\mathrm{SO}(3)$ on $\mathbb{S}^2$ is transitive and non-free.

$\square$

Remark As $\det(R) = -1$ for a reflection R and the determinant map $\det : \mathrm{O}(3) \to \{-1, 1\}$ is continuous, there is no possibility to continuously transform a rotation into a reflection. In that sense reflections are 'unphysical' rigid transformations (because, as $\det \mathbf{I} = 1$, the identity $\mathbf{I}$ is a rotation), and for this reason attention is often restricted to the set of rotations when material properties are considered (see e.g. the book of Bhattacharya [10]).

A (physical) *crystal* is an anisotropic and homogeneous body that provides a three-dimensional periodic composition of building blocks (atoms, ions, molecules). [17] Formally, crystals are classified according to the symmetries they provide. Colloquially, a crystallographic group is a group of transformations that forms the symmetry group of a discrete point lattice. In the ambient three-space, there are 230 groups that are distinguished in crystallography. There exists eleven pairs of enantiomorphic[9] *Laue groups* that are isomorphic (and therefore as abstract groups indistinguishable,

---

[9]i.e. mirror-symmetry related, such as left and right hands

cp. Def. 2.6), so that 219 abstract groups remain (see the International Tables of Crystallography A [22] for more detailed discussions and explanations). From this, the corresponding point groups, that are the groups of symmetries provided by a unit cell that generates the point lattice by translation, can be defined by using the fact that the translations form an (abelian) normal subgroup. Due to the assumptions in this text, this invariant subgroup can be identified with $\mathbb{R}^3$ (or $\mathbb{R}^n$ in a more general context). Formally, crystallographic groups are classified as the discrete subgroups of the group of Euclidean motions in three-space that are the symmetry groups of discrete point lattices, the so called Bravais lattices (see the books of Schwarzenbach [23] or Bhattacharya [10]). The next definition captures this formally.

Definition 2.21 (BRAVAIS LATTICES, CRYSTALLOGRAPHIC GROUPS AND POINT GROUPS) Let $B = \{b_1, \ldots, b_n\}$ be an orthonormal basis of $\mathbb{R}^n$ and $(\mathbf{o}, b_1, \ldots, b_n)$ a frame of reference. Elements in the frame of reference have the form

$$\mathcal{F}(\mathbf{o}, b_1, \ldots, b_n) = \{x \in \mathbb{R}^n | x = \sum_{i=1}^{n} a_i b_i + \mathbf{o} \text{ and } a_i \in \mathbb{R}\}.$$

For discrete lattices, the coefficients of the elements in $\mathcal{F}$ are be restricted to be integers

$$\mathcal{L}(\mathbf{o}, B) := \{x \in \mathbb{R}^n | x = \sum_{i=1}^{n} m_i b_i + \mathbf{o} \text{ and } m_i \in \mathbb{Z}\}.$$

Then, $\mathcal{L}(\mathbf{o}, B)$ is the *Bravais lattice* generated by B at $\mathbf{o}$. A discrete subgroup $\mathcal{C} < \mathrm{E}(n)$ is an *n-dimensional crystallographic group*, if there is a Bravais lattice $\mathcal{L}(B, \mathbf{o})$ with

$$\mathcal{C}(\mathcal{L}(\mathbf{o}, B)) = \mathcal{L}(\mathbf{o}, B).$$

If $n = 3$, the crystallographic groups are called *space groups*. Because $\mathbb{R}^n \trianglelefteq \mathcal{C}$, the factor group

$$\mathcal{P} := \mathcal{C}/\mathbb{R}^n$$

exists. This is called a *crystallographic point group.*
Two points $u, v \in \mathcal{L}(\mathbf{o}, B)$ are *crystallographically equivalent*, if there exists
a transformation $T \in \mathcal{C}$, such that $u = T(v)$.   □

Thus, point groups reflect the symmetries of finite objects (like a crystal's
unit cell) respecting a discrete point lattice, while crystallographic and
space groups describe (infinite) periodic structures (cp. [23]). Different
sets of linearly independent vectors centered at the same point $\mathbf{o}$ may
generate the same Bravais lattice. The next theorem states precisely the
equality of Bravais lattices generated by two sets of vectors (cp. [10]).

Theorem 2.10 (IDENTITY OF BRAVAIS LATTICES) Let $\{e_i | i = 1, 2, 3\}$,
$\{f_i | i = 1, 2, 3\} \subset \mathbb{R}^3$ be linear independent sets that form right-handed
systems. Let $\mathbf{o} \in \mathbb{R}^3$ and $\mathcal{L}(\mathbf{o}, \{e_i | i = 1, 2, 3\}$ and $\mathcal{L}(\mathbf{o}, \{f_i | i = 1, 2, 3\}$ be
Bravais lattices. Then

$$\mathcal{L}(\mathbf{o}, \{e_i | i = 1, 2, 3\}) = \mathcal{L}(\mathbf{o}, \{f_i | i = 1, 2, 3\}) \leftrightarrow$$

$$\exists T \in \mathrm{SL}(3) \cap \mathbb{Z}^{3 \times 3} : f_i = T e_i \text{ for } i = 1, 2, 3.$$

PROOF. The direction $\rightarrow$: If $\mathcal{L}(\mathbf{o}, \{e_i | i = 1, 2, 3\}) = \mathcal{L}(\mathbf{o}, \{f_i | i = 1, 2, 3\})$,
the lattices are indistinguishable. Then there is a transformation T that
relates e_i to f_i by $f_i = T e_i$. Because the lattices are oriented in the same
way, $\det T = 1$, so $T \in \mathrm{SL}(3)$. Because the point lattices are discrete,
$T \in \mathbb{Z}^{3 \times 3}$. /
The direction $\leftarrow$: Because $f_i = T e_i$ $(i = 1, 2, 3)$ and $T \in \mathrm{SL}(3) < \mathrm{GL}(3)$,
T is one-to-one and onto and does not alter distances or angles. So, the
Bravais lattices coincide. /
Thus, the proposition holds.   □

A rotation R in a space group is an *m-fold* rotation, if there is an $m \in \mathbb{N}$
with $R^m = \mathbf{I}$, and if m is minimal with this property. In space groups
such an m always exists. Further, it can be shown that $m \in \{1, 2, 3, 4, 6\}$
(cp. e.g. [23] or [17]), what restricts the number of possible space groups.
Arthur Schoenflies derived all possible 230 space groups by combining
all possible symmetry operations respecting point lattices in three-space

in [24], where the *Schoenflies notation* for space groups originates from.[10]
Following an earlier remark, the convention to exclude reflections from
the point groups is usually applied in this text and attention is restricted
to the subgroups of rotations (cp. also [10]).

An example shall illustrate the concept of point groups. The cubic point
group, denoted 432 in the Hermann-Mauguin notation (and O in the
Schoenflies notation[11]), appears frequently in this work.

Example 2.2 (POINT GROUP OF A CUBE) Assume a unit cube in three-
space to be given, i.e. the set $[0, 1]^3$. The cube has 24 rotation symmetries:

- The identity transformation.
- Six 2-fold rotations, see Fig. 2.1a.
- Four 3-fold rotations, see Fig. 2.1b.
- Three 4-fold rotations, see Fig. 2.1c.

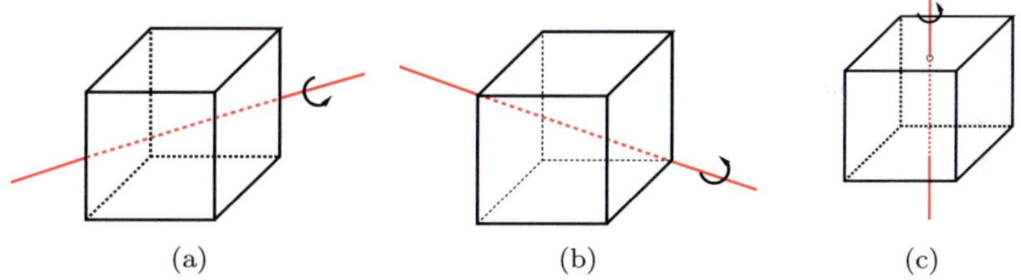

Figure 2.1: Non-equivalent rotation symmetries in a cube: **(a)** 2-fold axes,
(b) 3-fold axes and **(c)** 4-fold axes.

---

[10]Today, the *Hermann-Mauguin notation* is more widely used, because it is the
standard notation in the International Tables For Crystallography [22] to classify symmetries in crystals.

[11]O stands for the *octahedral group*. As cube and octahedron are dual platonic solids
(cp. the book of Coxeter [25]), every symmetry operation of a cube is a symmetry
operation of the octahedron and vice versa: The symmetry groups of octahedron and
cube are isomorphic.

Allowing mirror symmetries, the complete symmetry group has 48 elements, and is called the *full octahedral group*[12]. When one stretches the cube in two directions by the same magnitude and shortens it in the third, 4-fold symmetry axes at the faces are lost. The resulting symmetry group is a subgroup of the cubes' symmetry group containing 8 elements[13]. This loss of symmetry and the group-subgroup relationship is essential when it comes to a proper description of the symmetry-breaking martensitic transformation.   □

Crystallographic groups (and space groups) are groups that map an n-dimensional point lattice onto itself, and the point groups are the groups that leave at least one point on this lattice fixed. The number of (up to isomorphism) different possible crystallographic groups is finite for each $n \in \mathbb{N}$.[14] For $n = 3$, there are 14 possible Bravais lattices, classified by six *lattice constants*: the length of three linearly independent vectors $a, b, c \in \mathbb{R}^3$ and the three angles α, β, γ included between them (see Tab. 2.1). The Bravais lattice is generated by translation of a single unit cell spanned by a, b and c (see Fig 2.2a). When analyzing 'real' crystal structures, one can usually think of the points of the Bravais lattice as the mean positions of the vibrating atoms at a finite temperature (cp. [10]).

Crystal System	Vectors	Angles
Triclinic	$\|a\| \neq \|b\| \neq \|c\|$	$\alpha \neq \beta \neq \gamma$
Monoclinic	$\|a\| \neq \|b\| \neq \|c\|$	$\alpha = \gamma = 90°,\ \beta \neq 90°$
Orthorhombic	$\|a\| \neq \|b\| \neq \|c\|$	$\alpha = \beta = \gamma = 90°$
Tetragonal	$\|a\| = \|b\| \neq \|c\|$	$\alpha = \beta = \gamma = 90°$
Trigonal	$\|a\| = \|b\| \neq \|c\|$	$\alpha = \gamma = 90°,\ \beta = 120°$
Hexagonal	$\|a\| = \|b\| \neq \|c\|$	$\alpha = \gamma = 90°,\ \beta = 120°$
Cubic	$\|a\| = \|b\| = \|c\|$	$\alpha = \beta = \gamma = 90°$

Table 2.1: The seven crystal systems characterized by the six lattice constants.

---

[12]Written O_h in the Schoenflies and $m\bar{3}m$ in the short Hermann-Mauguin notation.

[13]Noted as D_4 in the Schoenflies and 422 in the short Hermann-Mauguin notation.

[14]This has been proven by Bieberbach in 1912 [26]. The proof is part of a more general solution to the 18th of the 23 problems of the Hilbert program proposed by David Hilbert in the 1920s.

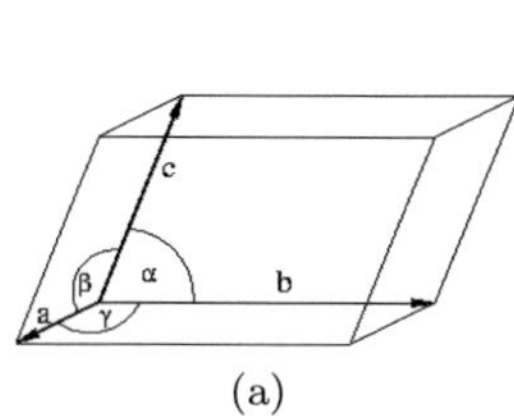
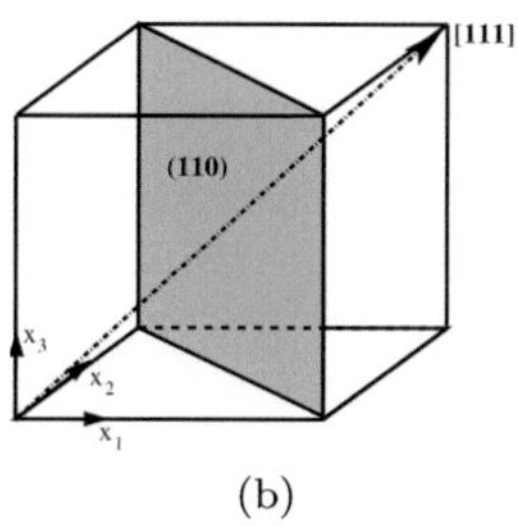

(a)   (b)

Figure 2.2: **(a)** A general unit cell showing the six lattice constants a, b, c and α, β, γ. **(b)** A simple cubic lattice with sketched crystallographic direction [111] (the cubes' diagonal) and plane (110).

The seven crystal systems can be ordered according to their respective group-subgroup relations, and thus form a 'mathematical lattice'[15]. Fig. 2.3 shows the lattice when the point groups are restricted to rotation operations. The full lattice is shown e.g. in the book of Borchert-Ott [17]. Directions and planes need to be identified in a Bravais lattice (e.g. to indicate the direction of shears and invariant planes to characterize martensite twins in shape memory alloys). The notation is restricted to three-dimensional spaces. To interpret the notation, the reader is referred to the definition of reciprocal bases (see Def. 2.15).

Definition 2.22 (DIRECTIONS AND PLANES IN BRAVAIS LATTICES) Let $\mathcal{L}(\mathbf{o}, B = \{b_1, b_2, b_3\})$ be a Bravais lattice in three-space. Let $u, v, w \in \mathbb{Z}$. The *crystallographic direction* $[u\ v\ w]$ is the vector

$$d = ub_1 + vb_2 + wb_3.$$

The set of all crystallographically equivalent directions is denoted by $\langle u\ v\ w \rangle$.
Let $h, k, l \in \mathbb{Z}$. The *plane in the Bravais lattice* denoted by $(h\ k\ l)$ is the plane with normal

$$n = hb^1 + kb^2 + lb^3,$$

---

[15] I.e. an ordered set. Cp. [27] for a proper definition.

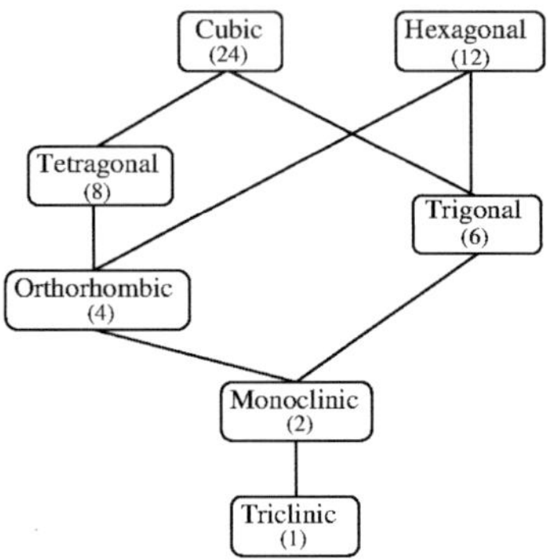

Figure 2.3: Relation between the seven crystal systems restricted to their rotation operations (in analogy to [10]). The graph shows the names of the systems and the number of rotation symmetries. Not all crystal systems are comparable, and two maximal elements (the cubic and the hexagonal system) exist.

given in the reciprocal basis. By $\{h\ k\ l\}$ the set of crystallographically equivalent planes is denoted. $\qquad\qquad\square$

In accordance with most literature that deals with crystallographic notation, the following notation is adopted:

Convention Negative numbers in directions and planes are denoted by a 'bar' atop the number instead of a minus sign, e.g. $[1\bar{1}1]$ instead of $[1-1\,1]$.

As an example, a cubic lattice is taken (again following [10]):

Example 2.3 (DIRECTIONS AND PLANES IN A CUBIC LATTICE) Assume a simple cubic lattice (see Fig 2.2b). The directions $[100], [010], [001]$ are the directions of the edges of the cube parallel to the basis vectors. Vectors parallel to the cubes' edges are crystallographically equivalent in the cubic point group, such that $\langle 100 \rangle = \{[100], [010], [001], [\bar{1}00], [0\bar{1}0], [00\bar{1}]\}$. In the simple cubic lattice, basis and reciprocal basis coincide, and all planes spanned by pairs of the basis vectors are crystallographically equivalent: $\{101\} = \{(110), (101), (011), (1\bar{1}0), (10\bar{1}), (01\bar{1})\}$. $\qquad\square$

2.4 Fourier transforms

The Fourier transform has many applications in the analysis of physical data and in numerical computation. For many problems solutions can be found in Fourier space, e.g. if derivations and convolutions of functions can are involved. *Fast Fourier transform* (FFT) techniques are a way to efficiently calculate Fourier transforms (see the textbook [28]), what makes these solutions efficiently computable. This is exploited to solve the arising equations in the micromagnetic problems efficiently (see Chap. 8). The field of complex numbers, $\mathbb{C}$ (see Ex. 2.1), is the image set of the functions under consideration. This section introduces the commonly used notation, whereas the idea of (Lebesgue-)integrable functions is assumed to be known. The definitions follow the book of Königsberger [29]. Again, let $n \in \mathbb{N}$ be fixed.

Definition 2.23 (INTEGRABLE AND DIFFERENTIABLE FUNCTIONS) A function $f : \mathbb{R}^n \to \mathbb{C}$ is an *integrable function*, if $\int_{\mathbb{R}^n} |f(x)|\, \mathrm{d}x < \infty$. The set of integrable functions is labeled $\mathcal{L}^1(\mathbb{R}^n)$. Let $m \in \mathbb{N}$. If the function f is m-times differentiable with respect to the variable x, then the m-th derivation is denoted by $f^{(m)} := \frac{\partial^m}{\partial x} f$. A function is said to be *smooth*, if f is arbitrary many times differentiable, abbreviated as $f \in \mathcal{L}^\infty(\mathbb{R}^n)$. $\square$

Now, Fourier transforms and convolutions are defined (cp. e.g. [29]).

Definition 2.24 (FOURIER TRANSFORMS AND CONVOLUTIONS) Let $f, g \in \mathcal{L}^1(\mathbb{R}^n)$. The *Fourier transform* of f is the function $\hat{f} : \mathbb{R}^n \to \mathbb{C}$ with

$$\forall x \in \mathbb{R}^n : \hat{f}(x) := \frac{1}{(2\pi)^{\frac{n}{2}}} \int_{\mathbb{R}^n} f(t) \exp(-i(x \cdot t))\, \mathrm{d}t,$$

where exp is the usual exponential function.

The integral

$$\forall x \in \mathbb{R}^n : (f \star g)(x) := \int_{\mathbb{R}^n} f(x-y)g(y)\,\mathrm{d}y$$

exists almost everywhere[16] in $\mathbb{R}^n$ and is called the *convolution* of f and g.

$\square$

The Fourier transforms (or later, when implementing algorithms on a computer, their discrete versions) will be used to significantly speed-up simulations and drop simulation times (actually making simulations possible in acceptable time). The next theorem states some important properties of Fourier transforms (cp. e.g. the book of Bracewell [28]).

Theorem 2.11 (PROPERTIES OF FOURIER TRANSFORMS) Let $f, g \in \mathcal{L}^1(\mathbb{R}^n)$. Then the following properties hold:

(i)

$$\widehat{f+g} = \hat{f} + \hat{g}. \qquad\qquad \text{(additivity theorem)}$$

(ii) Let $t \in \mathbb{R}^n$. Then

$$\hat{f}(x-t) = \exp(2\pi i t x)\hat{f}(x). \qquad\qquad \text{(translation theorem)}$$

(iii)

$$\widehat{f \star g} = (2\pi)^{\frac{n}{2}}\left(\hat{f} \cdot \hat{g}\right). \qquad\qquad \text{(convolution theorem)}$$

(iv) If $\int_{\mathbb{R}^n} |f|^m\,\mathrm{d}x < \infty$, then

---

[16]I.e. the set of points where the integral does not exist has Lebesque-measure zero.

$$\forall x \in \mathbb{R}^n : \hat{f}^{(m)}(x) = \exp(2\pi i x)^m \, \hat{f}(x). \qquad \text{(derivation theorem)}$$

$$\square$$

The *convolution theorem* states that the Fourier transform of the convolution of integrable functions f and g can be calculated as the point-wise product of the Fourier transforms of the functions. This theorem becomes important when the demagnetization field of a finite specimen has to be evaluated efficiently. The *derivation theorem*, on the other hand, is applied when the demagnetization field of a three-dimensional periodic specimen (i.e. representative volume element) has to be calculated. It enables to efficiently solve an arising *Laplace-type equation*[17].

2.5 Lie-group methods and exponentials

When solving partial differential equations (PDEs), it is convenient to choose the time-integration scheme adequately. The translation updates that usually occur in one-step Euler schemes can, for example, be replaced by schemes that use rotations if the integration is enforced to happen on a sphere. These schemes can be designed to be also explicit one-step schemes, but might avoid drawbacks. This section prepares the necessary framework for the solution scheme that is later used to compute the update for magnetic moments in a ferromagnetic body, which are, due to certain conditions, bound to evolve on the unit sphere $\mathbb{S}^2$. The field of geometric integration and Lie-group methods is a relatively new field in mathematics. Iserles et al. published a very well written introduction that motivates the need of geometric integration methods [30]. While the main idea can easily be paraphrased as: 'Choose the basic motions that solve your PDE at hand adequately for the problem', the underlying theory is rather complicated.

---

[17]A Laplace equation is a differential equation where the Laplace operator (cp. Def. 2.19) is involved, and with that second order spatial derivatives

The combination of groups and differentiability goes back to the works of Sofus Lie ($\star$ 1842 - $\dagger$ 1899) and is encountered in the concept of Lie-groups. Lie-groups are useful to study symmetries, invariants and the qualitative behavior of differential equations (see Iserles et al. [30]). Because the configuration manifold of a physical problem[18] is usually a non-linear space, it is not so easy to preserve the structure when solving a differential equation in this space. The basic idea is the following: Describe the problem as the action of a suitable Lie-group on this manifold, and solve it in an associated linear structure, the so called Lie-algebra. There, the differential equation is discretized using only linear operations that automatically preserve the linear structure. Then, the process is reversed to obtain a solution of the differential equation in the original manifold by exponentiation. Referring to a talk given by Iserles (see [31]), the basic ideas are summarized as follows: Let $\mathcal{M}$ be a differentiable manifold, G be a Lie-group that acts on $\mathcal{M}$, and let $\mathfrak{g}$ be the Lie-algebra associated to G. Let a set of differential equations evolving (in time) on $\mathcal{M}$ be given.

1. Transform the equations from $\mathcal{M}$ to G

2. Transform the equations from G to $\mathfrak{g}$

3. Discretize the equations in $\mathfrak{g}$ using only linear operations and solve them

4. Transform the result from $\mathfrak{g}$ to G

5. Transform the result from G to $\mathcal{M}$

By construction, the solutions gained in $\mathcal{M}$ meet all constraints.

For this section, definitions and notations strictly follow the book of Iserles et al. [30], as well as the cited theorems. An $n \in \mathbb{N}$ is again fixed. The discussion will be mostly restricted to matrix Lie-groups, but starts with the definition of special groups that reflect continuous symmetries.

---

[18]that is the set of allowed states for the problem

Definition 2.25 (Lie-Groups and Matrix Lie-Groups) Let $(G, \cdot)$ be a group. If G is a differentiable manifold[19] and the group multiplication and inversion,

$$(g_1, g_2) \mapsto g_1 \cdot g_2 \text{ and } g \mapsto g^{-1},$$

are smooth maps, then G is a *Lie-group*. If the elements of G are matrices, then G is a *matrix Lie-group*. □

Lie-algebras $\mathfrak{g}$ are vector spaces that provide a measure of commutativity of elements of $\mathfrak{g}$, called the commutator or Lie-bracket. As in this work the relevant Lie-groups are matrix Lie-groups and the correspondence between Lie-groups and Lie-algebras will be exploited, the next definition includes a special version of for matrix Lie-algebras.

Definition 2.26 (Lie-algebras and Matrix Lie-algebras) A *Lie-algebra* is an n-dimensional vector space V equipped with a bilinear map called *commutator map* or *Lie bracket* $[\cdot, \cdot\cdot] : V \times V \to V$, such that

$$\forall u, v \in V : [u, v] = -[v, u] \qquad \text{(skew symmetry)}$$
$$\forall u, v, w \in V : [u, [v, w]] = [w, [u, v]] = [v, [w, u]] \qquad \text{(Jacobi's identity)}$$

If $V = \mathbb{R}^{n \times n}$ and closed under the *matrix commutation* defined by

$$\forall A, B \in V : [A, B] = AB - BA,$$

then V is a *matrix Lie-algebra*, and $[\cdot, \cdot\cdot]$ is the *matrix commutator*. □

Further notation is necessary to describe the correspondence between Lie-groups and Lie-algebras. The text again follows strictly the book of Iserles et al. [30] Tangents at a point p of a manifold $\mathcal{M}$ will be introduced. The set of all tangents at p can be equipped with a vector space structure. This linear structure and its correspondence to a certain Lie-group will play a crucial role when solving PDEs.

---

[19]i.e. G is locally homeomorphic to $\mathbb{R}^m$ for some $m \in N$, cp. [29]

Definition 2.27 (TANGENT SPACES AND VECTOR FIELDS) Let $\mathcal{M}$ be a manifold, $p \in \mathcal{M}$ and $\rho : \mathbb{R} \to \mathcal{M}$ a (time-dependent) smooth curve with $\rho(0) = p$. Then the derivative of ρ at $t = 0$, written $\rho'(0)$, is a *tangent vector* at p. The set of all tangents through p is the tangent space $T\mathcal{M}|_p$ of p in $\mathcal{M}$. $T\mathcal{M}|p$ is a linear space.
The set of all tangent spaces at all points of $\mathcal{M}$ is

$$\Xi = \bigcup_{p \in \mathcal{M}} T\mathcal{M}|_p.$$

A (tangent) *vector field* is a function $F : \mathcal{M} \to \Xi$ such that

$$F(p) \in T\mathcal{M}|_p,$$

i.e. F associates to each point $p \in \mathcal{M}$ a tangent through p. The set of all possible vector fields is denoted by $\mathcal{X}(\mathcal{M})$ and again carries the structure of a vector space. $\qquad\square$

Many problems arising in physics and mechanics can be described by differential equations where the underlying configuration space has the structure of a manifold.

Definition 2.28 (DIFFERENTIAL EQUATIONS ON MANIFOLDS) Let $\mathcal{M}$ be a manifold and F a vector field on $\mathcal{M}$. A *differential equation* evolving on the manifold $\mathcal{M}$ is the problem of finding a function $y : \mathbb{R} \to \mathcal{M}$ that satisfy

$$y'(t) = F(y(t), t), \qquad t \in \mathbb{R}_{\geq 0} \qquad \text{and} \qquad y_0 := y(0) \in \mathcal{M}.$$

The *flow* produced by the vector field F is the operator $\Psi_{t,F}(y_0)$ with

$$y(t) = \Psi_{t,F}(y_0).$$

Given the flow $\Psi_{t,F}$, the vector field F can be found by differentiation of the flow:

$$F(y) = \frac{\mathrm{d}}{\mathrm{d}t}\Psi_{t,F}(y)|_{t=0}.$$

F is called the *infinitesimal generator* of the flow $\Psi_{t,F}$. $\qquad\square$

The relation

$$\Psi_{\alpha,F} = \Psi_{1,\alpha F}$$

is valid for all $\alpha \in \mathbb{R}$. This can be interpreted as reparametrizing time or rescaling the vector field (see [30]). The computation of flows is called *exponentiation* (see Ex. 2.4), written as $\Psi_{1,\alpha F} \equiv \exp(F)$ or equivalently as $\Psi_{t,\alpha F} \equiv \exp(tF)$.

The three-dimensional real space can be interpreted as the Lie-algebra to the matrix Lie-group of rotations of the real three-space. Two functions are introduced to relate these two algebraic structures by *algorithmic exponentials* that can be used to solve the differential equations for the evolution of micromagnetic moments in micromagnetic simulations: The matrix exponential map, defined in analogy to the exponential-function exp in $\mathbb{R}$, and the Cayley transform.

Definition 2.29 (Matrix Exponential and Cayley Transform) The *matrix exponential* is the map

$$\exp : \mathrm{GL}(n,\mathbb{R}) \to \mathrm{GL}(n,\mathbb{R}),$$

$$A \mapsto \sum_{i=0}^{\infty} \frac{1}{i!} A^i.$$

The *Cayley transform* is defined as

$$\mathrm{cay} : \mathrm{skew}(\mathbb{R}^{n \times n}) \to \mathrm{GL}(n,\mathbb{R}), A \mapsto (\mathbf{I} + \frac{1}{2}A)(\mathbf{I} - \frac{1}{2}A)^{-1}.$$

$\square$

The matrix exponential converges and maps invertible matrices to invertible matrices. The restriction of the function cay to skew symmetric matrices is explained in the examples at the end of this section, as well as the invertibility of matrices under cay. The following properties of the matrix exponential are needed later in this work. For proofs it is referred to the textbook of Hilgert and Neeb [32].

Theorem 2.12 (PROPERTIES OF THE MATRIX EXPONENTIAL) The matrix exponential exp has the following properties:

(i) For all commuting $X, Y \in \mathbb{R}^{n \times n}$ (i.e. matrices with $XY = YX$):

$$(\exp(X))(\exp(Y)) = \exp(X + Y).$$

(ii) For all $X \in \mathbb{R}^{n \times n}$

$$\exp(X^T) = \exp(X)^T.$$

(iii) $\det(\exp(X)) \neq 0$ for all $X \in \mathbb{R}^{n \times n}$, hence $\exp(X) \in \mathrm{GL}(n)$.

(iv) For all $X \in \mathbb{R}^{n \times n}$:

$$\det(\exp(X)) = e^{TrX},$$

where TrX denotes the *trace* of X, i.e. the sum of the diagonal elements of X.

$\square$

The matrix exponential occurs naturally in solving differential equations with functions that operate on real vector spaces. The example is taken from [30].

Example 2.4 (MATRIX DIFFERENTIAL EQUATIONS) Let L_A be a linear vector field on $\mathbb{R}^n$ given by $A \in \mathrm{GL}(n)$ via $L_A(y) = Ay$. Consider the differential equation

$$y'(t) = Ay(t), \qquad t \in \mathbb{R}_{\geq 0}, \qquad y_0 := y(0) = \in \mathbb{R}^n.$$

The solution is given by

$$y(t) = \exp(tA)y_0 = \sum_{j=0}^{\infty} \frac{1}{j!}(tA)^j y_0,$$

so the flow the solution produces is

$$\Psi_{t,L_A}(y_0) \equiv \exp(tL_A)(y_0) \equiv \exp(tA)y_0.$$

$\square$

If now G is a matrix Lie-group that acts on the manifold $\mathcal{M}$ via an operation Λ, and $\rho : \mathbb{R} \to G$ is a smooth curve satisfying $\rho(0) = \mathbf{I}$, then this curve produces a flow on $\mathcal{M}$, and by differentiation one gets a vector field F as

$$F(y) = \frac{\mathrm{d}}{\mathrm{d}t}\Lambda(\rho(t), y)|_{t=0}.$$

The collection of all such vector fields carries the structure of a Lie-algebra. To every Lie-group, a Lie-algebra can be associated by considering the set of all tangents to the identity element of the Lie-group.

The next theorem gives a main result for a correspondence between Lie-group elements and Lie-algebra elements by showing how for right-trivializable curves a direct solution for certain differential equations can be gained. The following proposition is taken directly from the book of Iserles et al. [30] and restricted to matrix Lie-groups (a proof for this and a generalized version is given there):

Theorem 2.13 (SOLUTION OF DIFFERENTIAL EQUATIONS ON MANIFOLDS) Let $\mathcal{M}$ be a manifold, and G be a matrix Lie-group that acts on $\mathcal{M}$ via Λ. Let $\mathfrak{g}$ be the associated Lie-algebra and $A \in \mathfrak{g}$. Set $\lambda_* : \mathfrak{g} \to \mathcal{X}(\mathcal{M})$ with

$$\lambda_*(a)(y) = \frac{\mathrm{d}}{\mathrm{d}t}\Lambda(\rho(t), y)|_{t=0},$$

where ρ is a curve in G with $\rho(0) = \mathbf{I}$ and $\rho'(0) = A$. Then

(i) λ_* is a Lie algebra homomorphism[20] from $\mathfrak{g}$ into $\mathcal{X}(\mathcal{M})$.

---

[20]I.e. λ_* respects the Lie-bracket.

(ii) The solution of the differential equation

$$y'(t) = \lambda_*(A)(y(t)) \qquad \text{for fixed} \qquad y(0) =: y_0 \in \mathcal{M}$$

can be expressed as

$$y(t) = \Lambda(S(t), y_0),$$

where $S : \mathbb{R} \to G, t \mapsto S(t)$ is a curve in G satisfying

$$S'(t) = AS(t), \quad t \geq 0, \quad S(0) = \mathbf{I}.$$

The explicit solution is given by

$$S(t) = \exp(tA), \qquad t \geq 0.$$

$\square$

The relation $S'(t) = AS(t)$ in Thm. 2.13 the *right trivialization* of the curve S: The (time) derivative of $S(t)$ is displayed as the matrix product of a Lie-algebra element and the original curve $S(t)$. This theorem will be used to construct an explicit one-step time integration scheme for the evolution of micromagnetic moments under certain boundary conditions, and follows the work of Lewis and Nigam (see [12] and Chap. 8). There are several ways to associate a Lie-algebra to a Lie-group. Usually, the concept is rather abstract and finding simple examples is difficult. The Lie-algebra to a Lie-group is the set of all tangents to the identity of the Lie-group. The matrix group of rotations in three space is identified with the group SO(3), and has the set of skew symmetric matrices as associated Lie-algebra. This can be seen as follows: Let $A(t)$ be a curve in SO(3) with $A(0) = \mathbf{I}$. Then, by applying the chain rule of differentiation,

$$\mathbf{0} = \frac{\mathrm{d}}{\mathrm{d}t} I = \frac{\mathrm{d}}{\mathrm{d}t} \left(A(t)A(t)^T \right) = A'(t)A(t)^T + A(t)(A'(t))^T,$$

and because $A(0) = \mathbf{I} = A^T(0)$

$$A'(t) = -(A'(t))^T,$$

i.e. the Lie-algebra elements $A'(0)$ are skew symmetric, and so $\mathfrak{so}(3) = \text{skew}(\mathbb{R}, 3)$. The Lie-bracket of a matrix Lie-algebra $\mathfrak{g}$ is the commutator map (cp. Def. 2.26)

$$[\cdot, \cdot\cdot] : \mathfrak{g} \times \mathfrak{g} \to \mathfrak{g}, (A, B) \mapsto [A, B] = AB - BA.$$

The introduced terminology shall be made more clear by giving an example.

Example 2.5 (THE LIE-ALGEBRA OF THE LIE-GROUP SO(2)) The set of rotations in 2D forms a Lie-group. This abstract group has the matrix representation SO(2) : Consider the counter-clockwise rotation about the angle $\alpha \in [0, 2\pi[$. Then

$$R_\alpha = \begin{pmatrix} \cos\alpha & -\sin\alpha \\ \sin\alpha & \cos\alpha \end{pmatrix}$$

is the matrix representation with the concatenation of matrices as the group multiplication. The map

$$R_\alpha \mapsto (\cos\alpha, \sin\alpha)^T$$

is an isomorphism between SO(2) and (a parametrization of) the unit circle $\mathbb{S}^1$. Let $\varphi : \mathbb{R} \to \text{SO}(2)$ be a path with $\varphi(0) = \mathbf{I}$. The map ρ defined by

$$\rho : \mathbb{R} \to \mathbb{S}^1, t \mapsto (\cos\varphi(t), \sin\varphi(t))^T$$

is a path on $\mathbb{S}^1$ with

$$\rho(0) = (1, 0)^T \quad \text{and} \quad \frac{d}{dt}\rho|_{t=0} = (-\sin\varphi(t), \cos\varphi(t))^T|_{t=0} = (0, 1)^T.$$

The set of all tangents at the identity, i.e. the Lie-algebra $\mathfrak{so}(2)$, is the line tangent to the identity $(1, 0)^T$ in the direction $(0, 1)^T$, so the Lie-algebra to the rotation group SO(2) is (isomorphic to) the real line. $\qquad\square$

One of the most important groups in this work, the group of rotations in 3D (identified with the matrix group SO(3)), has no simple visualization,

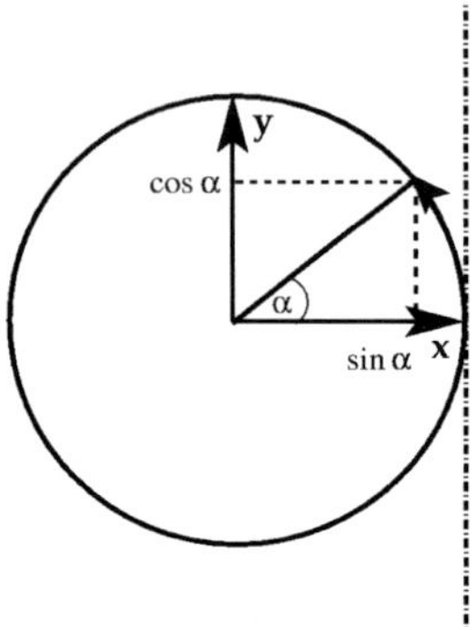

Figure 2.4: The unit circle $\mathbb{S}^1$ is a representation of SO(2), the Lie-algebra to SO(2) is set of tangents at the identity, and isomorphic to the real line. Unit circle and tangent space are sketched in the figure.

and in opposition to SO(2) the group SO(3) is non-abelian.[21] The set of so called unit quaternions (see the Appendix A.4) is a different representation of SO(3). The Lie-algebra $\mathfrak{so}(3)$ is related to the Lie-group SO(3) by the matrix exponential function:

Theorem 2.14 (SKEW SYMMETRIC MATRICES AND ROTATIONS) The matrix exponential exp maps matrices from $\mathfrak{so}(3)$ to rotations, i.e.

$$\exp : \mathfrak{so}(3) \to \mathrm{SO}(3).$$

PROOF. Let $B \in \mathfrak{so}(3)$ be a skew symmetric matrix. Thm. 2.12 is used to proof the defining properties of rotations (cp. Ex. 2.1) for $\exp(B)$: Because $B + B^T = B^T + B$

$$\begin{aligned}
\exp(B)\exp(B)^T &= \exp(B)\exp(B^T) \\
&= \exp(B + B^T) \\
&= \exp(B - B) \\
&= \exp(\mathbf{0}) \\
&= \mathbf{I}.
\end{aligned}$$

---

[21] In general, the groups SO(n) are abelian groups if and only if n is even.

Analogously, $\exp(B)^T \exp(B) = \mathbf{I}$ is shown, what proves $\exp(B)^T = \exp(B)^{-1}$.

Because the trace of the skew symmetric matrix B vanishes, it follows that

$$\det(B) = e^{TrB} = e^0 = 1.$$

So, $\exp(B) \in SO(3)$. $\qquad\qquad\qquad\qquad\qquad\qquad\qquad\qquad\qquad$ $\square$

The Lie-algebra $\mathfrak{so}(3)$ to $SO(3)$ can be identified with the space $\mathbb{R}^3$, equipped with the cross product, as will be shown now.

Definition 2.30 (THE MAP 'SKEW') The map defined by

$$\text{skew} : \mathbb{R}^3 \to \mathfrak{so}(3) = \text{skew}(\mathbb{R}, 3),$$

$$(x_1, x_2, x_3) \mapsto \begin{pmatrix} 0 & -x_3 & x_2 \\ x_3 & 0 & -x_1 \\ -x_2 & x_1 & 0 \end{pmatrix}.$$

is a bijection between the space $\mathbb{R}^3$ and the set of skew symmetric real 3×3-matrices. $\qquad\qquad\qquad\qquad\qquad\qquad\qquad\qquad\qquad\qquad$ $\square$

The map skew has important properties: It is a Lie-algebra homomorphism that identifies the real three-space with the set of skew-symmetric matrices, and it emerges naturally in the description of infinitesimal rotations.

Theorem 2.15 (PROPERTIES OF SKEW) Let $x, y \in \mathbb{R}$. Then:

(i) skew is a Lie-algebra isomorphism between the Lie-algebras $(\mathbb{R}^3, \times)$ and $(\mathfrak{so}(3), [\cdot, \cdot\cdot])$ (where the Lie-bracket of $\mathfrak{so}(3)$ is the matrix commutator map). Especially $x \times y = \text{skew}(x)y$.

(ii) Let R_{inf} be an infinitesimal rotation of a vector $v \in \mathbb{R}^3$, describing an infinitesimal change $v' = R_{\text{inf}}v$ of v. Then there is an infinitesimal $x \in \mathbb{R}^3$ such that $R_{\text{inf}} = (\mathbf{I} + \text{skew}(x))$.

PROOF.

(i) The proposition follows directly from the definition of the cross-product '$\times$' (cp. Def. 2.19) and the map skew. $\diagup$

(ii) Let e_1 and e_2 be two infinitesimal transformations. Then their concatenation $e_2 e_1$ is a negligible small transformation and

$$
\begin{aligned}
(\mathbf{I} + e_1)(\mathbf{I} + e_2) =& \mathbf{I}^2 + \mathbf{I}(e_1 + e_2) + e_1 e_2 \\
\approx& \mathbf{I} + e_1 + e_2 \\
=& \mathbf{I} + e_2 + e_1 \\
\approx& \mathbf{I}^2 + \mathbf{I}(e_2 + e_1) + e_2 e_1 \\
& (\mathbf{I} + e_2)(\mathbf{I} + e_1).
\end{aligned}
$$

i.e. $(\mathbf{I} + e_1)$ and $(\mathbf{I} + e_2)$ commute.
If e is an infinitesimal small transformation, then

$$
\begin{aligned}
(\mathbf{I} + e)(\mathbf{I} - e) =& \mathbf{I}^2 - e^2 \\
\approx& \mathbf{I},
\end{aligned}
$$

and

$$
\begin{aligned}
(\mathbf{I} - e)(\mathbf{I} + e) =& \mathbf{I}^2 - e^2 \\
\approx& \mathbf{I},
\end{aligned}
$$

i.e. $(\mathbf{I} + e)$ and $(\mathbf{I} - e)$ are inverse elements.
Let $R = (\mathbf{I} + e)$. If R is a rotation, then $R^{-1} = R^T$. Hence

$$
\begin{aligned}
R^T =& (\mathbf{I} + e)^T \\
=& \mathbf{I}^T + e^T \\
=& \mathbf{I} + e^T,
\end{aligned}
$$

what, in combination with $R^{-1} = (\mathbf{I} - e)$, gives

$$
e^T = -e.
$$

So, e is skew symmetric. Because e is infinitesimal small and skew symmetric, there is an infinitesimally small $x \in \mathbb{R}^3$ such that

$$e = \mathrm{skew}(x),$$

namely $x = (e_{32}, e_{13}, e_{21})^T$. ⁄

$\square$

The last two theorems show that to each real vector (or, equivalently, each skew symmetric matrix) a rotation matrix can be assigned. For skew symmetric matrices, the matrix exponential can be computed effectively. Let $A = (a_{ij}) \in \mathfrak{so}(3)$, and let $|A| := (\sum_{i,j} a_{ij}^2)^{\frac{1}{2}}$. The validity of the following equations is discussed in more detail by Iserles et al. [30]. The relation

$$A^3 = -|A|A,$$

holds, so that

$$\exp(A) = \mathbf{I} + \frac{\sin(|A|)}{|A|} A + \frac{1}{2} \frac{\sin^2(\frac{1}{2}|A|)}{(\frac{1}{2}|A|)^2} A^2.$$

The Cayley transform from Def. 2.29 relates skew symmetric matrices A to rotations in three-space in a very similar way. Because

$$(\mathbf{I} - \frac{1}{2}A)^{-1} = \mathbf{I} + \frac{1}{1 + |\frac{1}{2}A|} \left(\frac{1}{2}A + \left(\frac{1}{2}A \right)^2 \right)$$

$(\mathbf{I} - \frac{1}{2}A)^{-1}$ exists for all $A \in \mathrm{skew}(\mathbb{R}^{3\times3})$. It can be verified that $\mathrm{cay} : \mathfrak{so}(3) \rightarrow \mathrm{SO}(3)$. So, the Cayley transform is an alternative to the 'true' matrix exponential. The explicit formula

$$\mathrm{cay}(A) = \mathbf{I} + \frac{4}{4 + |A|^2} A + \frac{1}{2} \frac{4}{4 + |A|^2} A^2.$$

is valid and shows that $\mathrm{cay}(A)$ can be evaluated without the evaluation of trigonometric functions. Therefore, cay is often preferred over the true matrix exponential as an 'algorithmic exponential', because the evaluation of trigonometric functions can result in numerical inaccuracy.

Part II

Continuum Theories

3 Continuum mechanics

To model the behavior of a material accurately, phenomena occurring on different length scales and time scales have to be considered. The idea of *continuum mechanics* is to 'smear out' discrete events by the use of continuous field variables. The textbooks of Lai et al. [33] or Jaunzemis [34] give good introductions to the field of continuum mechanics, and a good brief review can be found in the book of Phillips [35, Chap. 2]. The present chapter deals with the continuum mechanics of solids and their deformation, a field called *kinematics*. The deformation of a solid body can be described by the vector field of displacement vectors that indicate the deformation from an initial reference configuration. From this, a relation between stress and strain states in a material, i.e. the relation between forces that act locally and the resulting macroscopic changes in length, can be defined. The link between the discrete events that take place in a material on a Bravais lattice and the continuum description is created by the Cauchy-Born hypothesis. For the models described in this work, the regime of small strains is assumed, and in addition, a linear theory of mechanics is applied.

3.1 Kinematics: Deformation and strain

For this section, let $\mathcal{B} \subset \mathbb{R}^3$ be an open, simply connected and bounded set that represents the solid under consideration. This section follows in its main parts the book of Phillips [35].

In discrete Newtonian mechanics, only a countable number of particles exists, which therefore can be labeled with integer numbers. This idea of labeling particles is transferred to the continuum by labeling the uncountably infinite number of positions with their spatial coordinates.[1] Assume a frame of reference $\mathcal{F} = \{\mathbf{o}, \{\mathbf{x}_1, \mathbf{x}_2, \mathbf{x}_3\}\}$ to be fixed. $\mathcal{B}$ is chosen as the reference state at fixed time $t = 0$. The position of a point at $X = (X_1, X_2, X_3)^T \in \mathcal{B}$ changes with time. At each specific time $t \in \mathbb{R}_{\geq 0}$ the position is given by the *deformation map*

$$x : \mathcal{B} \times \mathbb{R}_{\geq 0} \to \mathcal{F}, (X, t) \mapsto x(X, t) = (x_1(X, t), x_2(X, t), x_3(X, t))^T.$$

$x(\mathcal{B}, t)$ is the deformed body at time t, and $X = x(X, 0)$. The triples $(X_1, X_2, X_3)^T$ are called *material coordinates*. If the continuum $\mathcal{B}$ deforms, the description in the material coordinates X is called *material description* or *Lagrangian description*, while the description in terms of the deformation map x is called *spatial description* or *Eulerian description*. Usually, the deformation map is assumed to be injective for all relevant physical problems.[2] Furthermore, it is assumed that a description in the Lagrangian or the Euler description is equivalently possible.

The relation between $X \in \mathcal{B}$ in the reference configuration before a deformation and $x \in x(\mathcal{B})$ in the deformed state is given by the time-dependent displacement field

$$\mathbf{u} : \mathcal{B} \times \mathbb{R}_{\geq 0} : (X, t) \mapsto x(X, t) - X. \tag{3.1}$$

Measures for the deformation of $\mathcal{B}$ are based on the *deformation gradient* F. The deformation gradient quantifies the changes described by the deformation map x and has the nine components

$$F_{ij} = \frac{\partial x_i}{\partial X_j}, \qquad i, j = 1, 2, 3.$$

From Eq. (3.1) follows

$$x(X, t) = X + \mathbf{u}(X, t),$$

so

$$F = \mathbf{I} + \nabla \mathbf{u}$$

---

[1] One can arrive at a similar description when the idea of particles is abandoned, and only the continuous space is considered.

[2] This means that no interpenetration of the body with itself is allowed, and that a body cannot be shrunken to a single point.

or

$$F_{ij} = \delta_{ij} + \frac{\partial u_i}{\partial X_j}, \qquad i,j = 1,2,3.$$

in indicial notation.[3] The different deformation measures are motivated by considerations about how (infinitesimal) line segments, areas and volume segments deform. As the deformation takes place according to the deformation gradient F, the following relations hold (see [36]):

Theorem 3.1 (DEFORMATION OF DIFFERENTIAL LINES, AREAS AND VOLUMES) Let $p \in \mathcal{B}$ be a material point in the reference configuration that deforms according to a deformation gradient F. Let dL, dA and dV be an oriented differential line segment, area segment and volume segment at p in the reference configuration, and dl, da and dv the corresponding segments in the deformed configuration. By $\hat{n}$ the unit normal to dA and by $\hat{m}$ the unit normal to da are denoted (both pointing outwards). Then

(i) $dl = FdL$

(ii) $da = s \, \mathrm{cof} F dA$

(iii) $dv = s \det F dV$

(iv) $\hat{m} = |\mathrm{cof} F \hat{n}|^{-1}(\mathrm{cof} F \hat{n})$

$s \in \{-1, 1\}$ is the sign of $\det F$ and chosen in a way that the normal to dv points outwards. $\mathrm{cof} F$ is the *cofactor matrix* of F.[4] $\square$

Assuming two neighboring material points to be separated in the reference configuration by the infinitesimal vector segment dX with length dL that transforms into dx with length dl, one obtains, using $dx = FdX$ (see Th. 3.1), the expression

$$dl^2 - dL^2 = (dx \cdot dx) - (dX \cdot dX)$$

---

[3]The *Kronecker Delta* δ_{ij} is defined as $\delta_{ij} = \begin{cases} 1 & \text{if } i = j \\ 0 & \text{if } i \neq j \end{cases}$.

[4]The cofactor matrix of a matrix $M \in \mathbb{R}^{3 \times 3}$ is defined as follows: Let M^{ij} be the 2×2 matrix gained from M by deleting the i-th row and the j-th column. Then the (i,j)-th entry of the cofactor matrix $\mathrm{cof} M$ is defined as $(\mathrm{cof} M)_{ij} = (-1)^{i+j} \det M^{ij}$.

$$\begin{aligned}
&= (F\mathrm{d}X \cdot F\mathrm{d}X) - (\mathrm{d}X \cdot \mathrm{d}X) \\
&= (F\mathrm{d}X)^T (F\mathrm{d}X) - (\mathrm{d}X \cdot \mathrm{d}X) \\
&= \mathrm{d}X^T F^T (F\mathrm{d}X) - (\mathrm{d}X \cdot \mathrm{d}X) \\
&= \mathrm{d}X \cdot (F^T F\mathrm{d}X) - (\mathrm{d}X \cdot \mathrm{d}X) \\
&= \mathrm{d}X \cdot (F^T F - \mathbf{I}) \cdot \mathrm{d}X \\
&= \mathrm{d}X \cdot 2E \cdot \mathrm{d}X
\end{aligned}$$

with

$$E = \frac{1}{2}(F^T F - \mathbf{I}). \tag{3.2}$$

E is the commonly used *Lagrangian strain tensor*. Using $F = \mathbf{I} + \nabla\mathbf{u}$, the components of E are

$$E_{ij} = \frac{1}{2}\left(\frac{\partial u_i}{\partial X_j} + \frac{\partial u_j}{\partial X_i} + \sum_{k=1}^{3} \frac{\partial u_k^2}{\partial X_j \partial X_i} \right), \qquad i,j = 1,2,3.$$

As an example, a simple shear deformation will be analyzed (cp. [35]).

Example 3.1 (A SIMPLE SHEAR DEFORMATION) Assume a simple shear deformation in $\mathbb{R}^3$, that is an isochoric plane deformation in a direction s with magnitude γ in a plane E with normal n. The shear deformation leaves points on the plane E fixed, and points that lie outside the plane E are moved parallel to the plane in the direction of s with a magnitude proportional to the points distance to the plane and γ (see Fig. 3.1). For the plane E and the line with direction s the relations $E = \ker(s \otimes n)$ and $\{\alpha s \mid \alpha \in \mathbb{R}\} = (s \otimes n)(\mathbb{R}^3)$ are valid (cp. Def. 2.16). Because of $s, n \in \mathbb{S}^2$, the deformation gradient can be written as

$$F = \mathbf{I} + \gamma(s \otimes n).$$

In the concrete case where $s = (1,0,0)^T$, $n = (0,0,1)^T$ and $\gamma \in \mathbb{R}$ (cp. [35]), i.e. where the shear movement is of the amount γ in the $\mathbf{x}_1$-direction parallel to on the $(\mathbf{x}_2, \mathbf{x}_3)$-plane, the deformation mapping x is given by

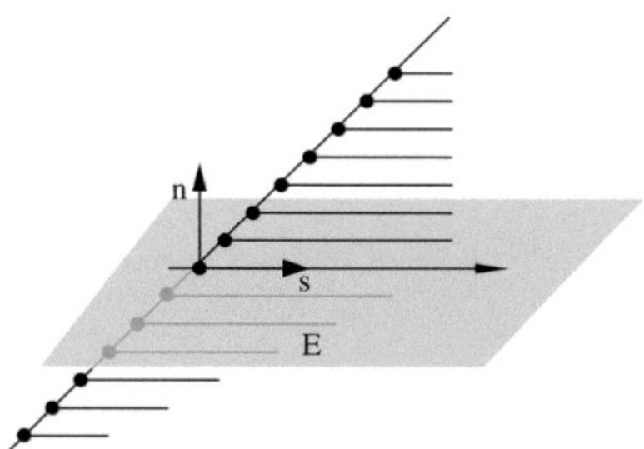

Figure 3.1: Illustration of a shear deformation in direction of s on a plane E with normal n: The plane of shear E is shown and the shear movement of points that lied on line along the direction n in the reference configuration. The shear direction is parallel to the plane E.

$$(x_1, x_2, x_3)^T \overset{x}{\mapsto} (X_1 + \gamma X_3, X_2, X_3)^T,$$

and the deformation gradient F has the form

$$F = \mathbf{I} + \gamma \left((1,0,0)^T \otimes (0,0,1)^T\right) = \begin{pmatrix} 1 & 0 & \gamma \\ 0 & 1 & 0 \\ 0 & 0 & 1 \end{pmatrix}.$$

$\square$

From Thm. 3.1, the so called *kinematic compatibility condition* can be derived (see [37]). This relation points out how two parts of a body Ω behave when they are subject to homogeneous deformations. If the deformations are compatible, then there has to be a plane separating the two parts, on which the deformation gradients act identically. This is also called the *invariant plane condition* or *Hadamard Jump Condition*.

Theorem 3.2 (HADAMARD JUMP CONDITION) Let $\Omega \subset \mathbb{R}^3$ be a mechanical body, and let $\Omega_1, \Omega_2 \subseteq \Omega$ with $\Omega = \Omega_1 \cup \Omega_2$ and $\Omega_1 \cap \Omega_2 = E$, where

E is a plane with normal $\hat{n}$ pointing from Ω_1 to Ω_2. Think Ω to be subject to a deformation with deformation gradient satisfying

$$F = \begin{cases} F_1 + c_1 & \text{if } x \in \Omega_1 \\ F_2 + c_2 & \text{if } x \in \Omega_2 \end{cases}.$$

Then

$$\text{cof}F_1\hat{n} = \text{cof}F_2\hat{n}.$$

$\square$

An equivalent formulation of the above theorem is that two deformation gradients F_1 and F_2 satisfy the kinematic compatibility if and only if (cp. [36] and remind the definition Def. 2.16)

$$F_1 - F_2 = a \otimes \hat{n} \tag{3.3}$$

for some vectors $a \in \mathbb{R}^3$ and $\hat{n} \in \mathbb{S}^2$. From this relation it is clear that $F_1 - F_2$ is of rank 1 as $\ker(F_1 - F_2)$ has dimension 2. Therefore, the kinematic compatibility condition is also called a *rank-one compatibility condition*.

3.2 Cauchy-Born hypothesis and free energies

The discrete structure of a material on the atomic scale as described by Bravais lattices (cp. Def. 2.21 and Sec. 2.3) is related to the continuum theory introduced in the last section via the so called *Cauchy-Born hypothesis*. The arguments in this section follow the book of Bhattacharya [10]. Let $\Omega \subset \mathbb{R}^3$ be a continuous body. The key idea is to attach a Bravais lattice $\mathcal{L}(x, \{e_i^0(x)|i = 1, 2, 3\})$ to each point $x \in \Omega$ of the continuum. Under an applied deformation with deformation gradient F, the Cauchy-Born hypothesis states that the lattice vectors deform according to F, so that for the lattice $\mathcal{L}(x, \{e_i(x)|i = 1, 2, 3\})$ at the same point x after the deformation

$$e_i(x) = F(x)e_i^0(x) \quad \text{for } i = 1, 2, 3$$

holds. To visualize this procedure, following [10], one can think of zooming into the structure using a high resolution microscope (this is illustrated in Fig. 3.2). In this text, the undeformed reference configuration is homogeneous[5], so that the Bravias lattices have no spatial dependence. Some choices for lattice vectors generate the same lattices. If $\{e_i | i = 1, 2, 3\}$ and $\{f_i | i = 1, 2, 3\}$ generate two Bravais lattices at the same point $x \in \mathbb{R}^3$, one lattice can be interpreted as a deformation of the other (see Thm. 2.10).

The Cauchy-Born hypothesis is applied to define free energy expressions that depend solely on the deformation gradient. If ψ is a function that assigns a Bravais lattice to each point of Ω,[6] then the temperature-dependent (Helmholtz) free energy $\hat{\varphi} : \psi(\Omega) \times \mathbb{R}$ of the system under consideration needs to satisfy two important conditions (cp. [10] and [38, 39]):

1. Frame-indifference: For all rotations $Q \in \mathrm{SO}(3)$

$$\hat{\varphi}(\{Qe_i^0\}, T) = \hat{\varphi}(\{e_i^0\}, T).$$

2. Material symmetry: For all $H \in \mathrm{SL}(3) \cap \mathbb{Z}^{3 \times 3}$:

$$\hat{\varphi}(\{He_i^0\}, T) = \hat{\varphi}(\{e_i^0\}, T).$$

The principle of *frame-indifference* states that the energy is independent of the position of an observer, i.e. the invariance under any change of

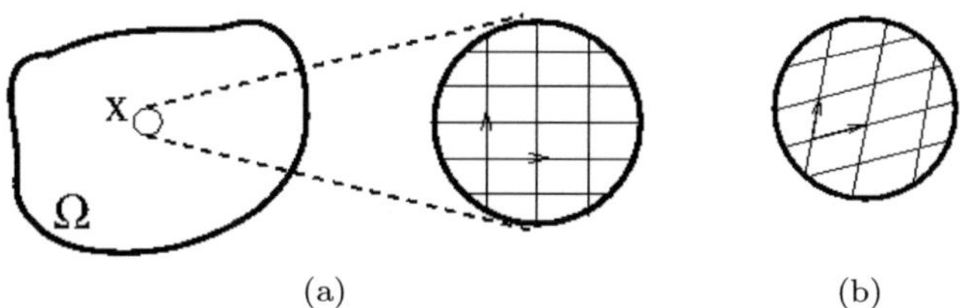

Figure 3.2: Illustration of the Cauchy-Born hypothesis: An elastic body Ω with Bravais lattice attached to a point $X \in \Omega$ **(a)** before and **(b)** after deformation. The figure is analogous to a figure shown in [10].

---

[5]In this work, the reference configuration is usually a cubic (austenite) state.

[6]This is, as the choice of lattice vectors is not unique, an 'act of choice' (cp. the footnote on page 7).

the frame of reference, while *material symmetry* demands that equivalent lattices have the same energy, i.e. that the energy is invariant of the choice of vectors generating the lattice (cp. Thm. 2.10).

Now, the Cauchy-Born hypothesis is used to link the energy defined as a function of the lattice and the temperature T to an energy expression depending on the deformation gradient describing the deformation. A homogeneous reference configuration is assumed. Let $\mathcal{L}(x, \{e_i\})$ be a fixed Bravais lattice at a point $x \in \Omega$. Then, the free energy is given by $\hat{\varphi}(\{e_i^0\}, T)$. As, according to the Cauchy-Born hypothesis, the lattice transforms according to deformation gradient at x, the free energy in terms of the deformation gradient is defined to be

$$\varphi(F, T) := \hat{\varphi}(\{Fe_i^0 | i = 1, 2, 3\}, T).$$

Let $\mathcal{P}$ denote the point-group of a Bravais lattice $\{x, \{e_i^0 | i = 1, 2, 3\}\}$. This energy needs to fulfill two essential requirements:

1. Frame-indifference: For all rotations $Q \in \mathrm{SO}(3)$

$$\varphi(QF, T) = \varphi(F, T).$$

2. Material symmetry: For all rotations $R \in \mathcal{P}$:

$$\varphi(FR, T) = \varphi(F, T).$$

These principles are the continuous versions of the invariance criteria for energies defined on discrete lattices.

3.3 Mechanical stress and strain

This section follows the book of Phillips [35, Sec. 2.3]. Let $\Omega \subset \mathbb{R}^3$ be a continuous mechanical body, and $\partial\Omega$ denote its boundary. Mechanical bodies experience forces either as *body forces* (such as gravity), or via *surface tractions* (e.g. by deforming them, or moving them around in space) [35]. The net force the body experiences is expressed as (cp. [35])

$$\int_{\Omega} f(r) \, \mathrm{d}\Omega + \int_{\partial\Omega} t(r) \, \mathrm{d}\partial\Omega, \tag{3.4}$$

where $f(r)$ is the force per unit volume in the point $r \in \Omega$ and $t(r)$ the force per unit area in the point $r \in \partial\Omega$. The first addend in Eq. (3.4) reflects the body forces, the second the traction forces. From this, by assuming that each physical system tends towards an equilibrium state of lowest energy, the existence of a (space and time-dependent) tensor quantity describing the stress state of a mechanical body can be derived (known as the Cauchy stress principle), as well as the dynamic equations of the continuum that describe how the system moves toward equilibrium.

3.3.1 The Cauchy stress principle

The *Cauchy stress principle* results from considerations about the equilibrium of body forces and traction forces that act on a (elementary) tetrahedron with three planes with normals n_1, n_2 and n_3 parallel to a Cartesian coordinate system $(\mathbf{o}, \{e_i | i = 1, 2, 3\})$ (see Fig. 3.3). The traction t^n on the plane with normal n that meets the three other planes can be determined, if the tractions

$$t^{e_i} = \sum_{j=1}^{3} \sigma_{ji} e_j.$$

on the three perpendicular with normals n_i planes are known. Gathering the components σ_{ji} in a tensor $\boldsymbol{\sigma}$, then

$$t^n = \boldsymbol{\sigma} n$$

can be proven, if the equilibrium of the elementary volume element is assumed. This is known as the *Cauchy stress theorem*:

Theorem 3.3 (Cauchy Stress Theorem) Let $\Omega \subset \mathbb{R}^3$ be a mechanical body. Then there is a tensor $\boldsymbol{\sigma}$, such that the traction acting on an arbitrary point (thought of as the limit of an infinitesimal area) at the surface $x \in \partial\Omega$ with surface normal $n(x)$ is given by

$$t^n(x) = \boldsymbol{\sigma}(x)n(x).$$

$\square$

The theorem states the existence of a tensor field that locally describes the stress state in each point x of the body, while the tensor $\boldsymbol{\sigma}$ itself is independent of the normal at x.

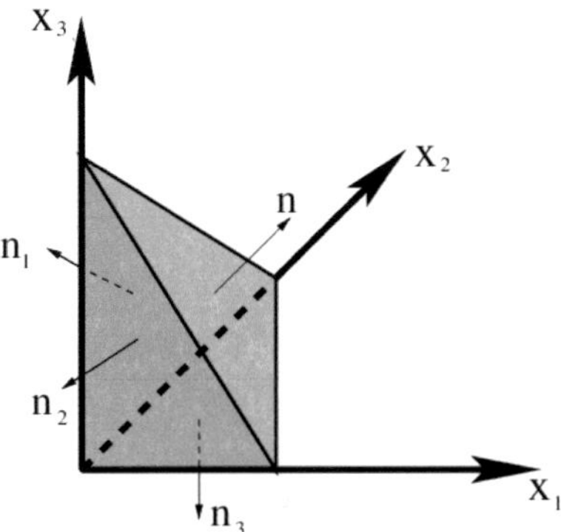

Figure 3.3: Illustration of the Cauchy stress principle (following [35]): The tetrahedral segment, a plane with normal n and the normals n_i of the segments the plane encloses with the Cartesian coordinate planes.

3.3.2 Continuum equations of motion

To gain the equations of motion in a continuous body, let $\bar{\Omega} \subset \Omega$ be an arbitrary subregion of a mechanical body Ω. The linear momentum is defined as follows (cp. [35]):

Definition 3.1 (LINEAR MOMENTUM) Let $\rho : \bar{\Omega} \to \mathbb{R}$ be the density function for $\bar{\Omega}$, and $v : \bar{\Omega} \times \mathbb{R}_{\geq 0} \to \mathbb{R}^3$ a time-dependent velocity field. Then

$$P := \int_{\bar{\Omega}} \rho v \, d\bar{\Omega}$$

is the *linear momentum*. $\square$

More notation is required to derive a continuous version of the law that states the equality of force as mass times acceleration[7]. This section again strictly follows the book of Phillips [35].

Definition 3.2 (MATERIAL TIME DERIVATIVE) Let $v : \bar{\Omega} \times \mathbb{R}_{\geq 0} \to \mathbb{R}^3$ be a time-dependent velocity field. The *material time derivative* is defined as the differential operator

$$\frac{\mathrm{D}}{\mathrm{D}t} = \frac{\partial}{\partial t} + v \cdot \nabla.$$

The velocity field is related to the displacement field via $v = \frac{\partial \mathbf{u}}{\partial t}$. $\qquad \square$

With this, the continuum version of the conservation of linear momentum becomes

$$\int_{\bar{\Omega}} f \, \mathrm{d}\bar{\Omega} + \int_{\partial\bar{\Omega}} t \, \mathrm{d}\partial\bar{\Omega} = \frac{\mathrm{D}}{\mathrm{D}t} \int_{\bar{\Omega}} \rho v \, \mathrm{d}\bar{\Omega}, \qquad (3.5)$$

where $\rho : \bar{\Omega} \to \mathbb{R}$ is the density, and f and t are the body forces and tractions acting on the body. The assumption of the conservation of mass reads

$$0 = \frac{\mathrm{D}}{\mathrm{D}t}\rho = \frac{\partial}{\partial t}\rho + \nabla \cdot (\rho v).$$

From this follows by application of the Reynolds transport theorem Thm. 2.8

$$\frac{\mathrm{D}}{\mathrm{D}t} \int_{\bar{\Omega}} \rho v \, \mathrm{d}\bar{\Omega} = \int_{\bar{\Omega}} \rho \frac{\mathrm{D}}{\mathrm{D}t} v \, \mathrm{d}\bar{\Omega}. \qquad (3.6)$$

From Cauchy's theorem Thm. 3.3, the surface traction can be written in terms of a stress tensor $\boldsymbol{\sigma}$, such that $t^n = \boldsymbol{\sigma}n$, and from the divergence theorem Thm. 2.7 follows

$$\int_{\bar{\Omega}} t(r) \, \mathrm{d}\bar{\Omega} = \int_{\bar{\Omega}} \boldsymbol{\sigma}n \, \mathrm{d}\partial\bar{\Omega} = \int_{\bar{\Omega}} \nabla \cdot \boldsymbol{\sigma} \, \mathrm{d}\bar{\Omega}.$$

---

[7]This is Newtons second law, abbreviated as $F = ma$ (see the book of Kibble and Bershire [40]).

With that and Eqs. (3.5) and (3.6) follows

$$\int_{\bar\Omega} \left(\nabla \cdot \boldsymbol{\sigma} + f - \rho \frac{\mathrm{D}}{\mathrm{D}t} v \right) \, \mathrm{d}\bar\Omega = 0. \tag{3.7}$$

As $\bar\Omega$ was an arbitrarily chosen subregion, the integrand of Eq. (3.7) has to vanish, giving the conservation law

$$\nabla \cdot \boldsymbol{\sigma} + f = \rho \frac{\mathrm{D}}{\mathrm{D}t} v.$$

The *mechanical equilibrium* in the absence of body forces is then expressed as

$$\nabla \cdot \boldsymbol{\sigma} = \mathbf{0}, \tag{3.8}$$

i.e. a system is in mechanical equilibrium if the tensor field of stresses, $\boldsymbol{\sigma}$, is divergence-free. The *elastic energy* of the mechanical body Ω is defined in terms of the stress tensor and the Lagrangian strain tensor as

$$E_{\mathrm{elast}} = \int_{\Omega} f_{\mathrm{elast}} \, \mathrm{d}\Omega = \frac{1}{2} \int_{\Omega} \boldsymbol{\sigma} \cdot E \, \mathrm{d}\Omega, \tag{3.9}$$

with the *elastic energy density* $f_{\mathrm{elast}} = \frac{1}{2}\boldsymbol{\sigma} \cdot E$.

3.4 Linear elasticity

This section introduces the linear theory of elasticity. It has to be differentiated between two kinds of linear elasticity: a *physically linear theory of elasticity* that states a linear relation between stress and strain as an approximation of material behavior, and a *geometrically linear theory of elasticity* that linearizes the elastic strain. The first is widely accepted, because the physically non-linear theory is assumed to exhibit no significant advantages in the prediction of material behavior (see e.g. [41]). There is no such agreement on the geometrically linear theory of elasticity. It is argued in the literature that in certain cases the geometrically linear theory of elasticity is insufficient to gain appropriate results (cp. e.g. [42]).

3.4.1 Geometrical linearization of the mechanical theory

If the gradients of the displacement field $\mathbf{u} : \Omega \times \mathbb{R}_{\geq 0}$ are 'small' in the sense of $|\frac{\partial u_i}{\partial X_j}| \ll 1$ for all $i, j = 1, 2, 3$, the last addend in the Lagrangian strain tensor E that contains higher order terms (see Eq. (3.2)) can be neglected. The result is the *infinitesimal strain tensor* or *small strain tensor* that reads

$$\epsilon_{ij} = \frac{1}{2} \left(\frac{\partial u_i}{\partial X_j} + \frac{\partial u_j}{\partial X_i} \right) \qquad \text{or} \qquad \epsilon = \frac{1}{2} \left(\nabla \mathbf{u} + \nabla \mathbf{u}^T \right).$$

This geometrically linearized version of the strain measure is a sufficient approximation in many physical cases. The definition directly shows that the linear strain tensor is a symmetric tensor.

In the geometrically linearized theory of kinematics, a version of the polar decomposition theorem can be formulated (see [10]):

Theorem 3.4 (Polar Decomposition: Linear Version) Let $H = \nabla \mathbf{u}$. Then there are $W \in \text{skew}(\mathbb{R}^{3 \times 3})$ and $E \in \text{symm}(\mathbb{R}^{3 \times 3})$, such that

$$H = W + E.$$

Proof. With

$$E = \frac{1}{2} \left(\nabla \mathbf{u} + \nabla \mathbf{u}^T \right) \text{ and } W = \frac{1}{2} \left(\nabla \mathbf{u} - \nabla \mathbf{u}^T \right)$$

the proposition holds. $\qquad \square$

For small $\mathbf{u}$, W is the skew-symmetric *(infinitesimal) rotation matrix* (cp. Th. 2.15). In this linearized theory the energy only depends on the displacement gradient (instead of on the deformation gradient in the nonlinear theory). The following version of frame indifference and material symmetry holds (where again $\mathcal{P}$ is the point group of the material):

1. Frame indifference: For all (infinitesimal) rotations $W \in \text{skew}(\mathbb{R}^{3 \times 3})$

$$\varphi(H + W, T) = \varphi(H, T).$$

2. Material symmetry: For all rotations $R \in \mathcal{P}$:

$$\varphi(R^T H R, T) = \varphi(H, T).$$

The second requirement can easily be shown to be equivalent to

$$\varphi(R^T E R, T) = \varphi(E, T)$$

for all rotations $R \in \mathcal{P}$. Having two deformation gradients $F_1 = \mathbf{I} + H_1$ and $F_2 = \mathbf{I} + H_2$, the kinematic compatibility reads

$$a \otimes n = F_1 - F_2 = H_1 - H_2$$

for some $(a, n) \in \mathbb{R}^3 \times \mathbb{S}^2$. The linear version of the polar decomposition theorem gives

$$H_1 = E_1 + W_1 \qquad \text{and } H_2 = E_2 + W_2,$$

such that

$$E_1 - E_2 = \frac{1}{2} \left(a \otimes n + n \otimes a \right) \tag{3.10}$$

and

$$W_1 - W_2 = \frac{1}{2} \left(a \otimes n - n \otimes a \right).$$

Eq. (3.10) is known as the *strain compatibility equation* (see [10]).

3.4.2 Physical linear theory of elasticity and Hooke's law

Here, the linear strains as introduced in the last section are assumed. From the work of Hooke in the 1660s, the assumption is adopted that a material responds linearly to external strains. Following [35], assume a simple one-dimensional setting: the unit force F (in Newton) acting on an area A (in square meter) is linearly related to the relative change in length ($\frac{\Delta l}{l}$, with l length in meter before, and Δl the length after the deformation):

$$\frac{F}{A} = E \frac{\Delta l}{l}.$$

The proportionality constant E is the *elastic modulus* or *Young's modulus* (having units of Newton per square meter, or Pascal). In a general three-dimensional setting, the stress is a tensor quantity $\sigma \in \mathbb{R}^{3\times 3}$, and its components σ_{ij} are linearly related to all nine components of the linear strain tensor ϵ as

$$\sigma_{ij} = \sum_{k,l=1}^{3} C_{ijkl}\epsilon_{kl}. \tag{3.11}$$

This is the three-dimensional version *Hooke's law of elasticity*. The proportionality tensor $\mathcal{C} = (C_{ijkl})$ is the *elastic stiffness* or *elastic modulus tensor*, and provides the elastic information about a linear elastic material under mechanical load (such as stiffness and symmetry). Assuming Hooke's law, the elastic energy Eq. (3.9) reads (using Einsteins summation convention)

$$E_{\text{elast}} = \int_{\Omega} \frac{1}{2}\epsilon \cdot \mathcal{C}\epsilon \, \mathrm{d}\Omega = \frac{1}{2}\int_{\Omega} C_{ijkl}\epsilon_{ij}\epsilon_{kl} \, \mathrm{d}\Omega \tag{3.12}$$

The elastic stiffness tensor $\mathcal{C}$ relates stresses and strains uniquely, so it is invertible. Its inverse $\mathcal{S}$ is the *elastic compliance tensor*. With the invertibility of $\mathcal{C}$, from Eq. (3.11) directly follows that a completely stress-free state $\sigma = \mathbf{0}$ is equivalent to the completely unstrained state $\epsilon = \mathbf{0}$, as the invertibility of $\mathcal{C}$ implies $\ker(\mathcal{C}) = \{\mathbf{0}\}$. The number of independent components decreases with increasing material symmetry. The elastic stiffness tensor $\mathcal{C}$ reflects the point group symmetry provided by the material.

From Eq. (3.12) follows

$$\sigma_{ij} = \frac{\partial E_{\text{elast}}}{\partial \epsilon_{ij}} \qquad \text{and} \qquad C_{ijkl} = \frac{\partial^2 E_{\text{elast}}}{\partial \epsilon_{ij}\partial \epsilon_{kl}},$$

and because $\sigma, \epsilon \in \text{symm}(\mathbb{R}^{3\times 3})$, of the 81 components of $\mathcal{C}$ only 21 can be independent (cp. [35]).

3.5 The concept of eigenstrains

A material may exhibit so called *eigenstrains* ϵ_0, when it is strained with respect to a reference state, but does not exert any stress. That is why eigenstrains are also called *stress-free strains*. Eigenstrains arise for example during the solidification of a material when precipitates form, or during the martensitic transformation as it will be described in Chap. 5. Detailed descriptions of eigenstrains and the method of eigenstrains can be read in the books of Phillips [35] and Gross and Seelig [43].

To account correctly for the eigenstrain ϵ_0 of a material, the elastic strain in the elastic energy equation has to be corrected by the influences from these eigenstrains. The elastic energy Eq. (3.9) becomes

$$E_{\text{elast}} = \int_\Omega \frac{1}{2}(\epsilon - \epsilon_0)\cdot\sigma \, d\Omega, = \int_\Omega \frac{1}{2}(\epsilon - \epsilon_0)\cdot\mathcal{C}(\epsilon - \epsilon_0) \, d\Omega, \quad (3.13)$$

with the elastic stress

$$\sigma = \mathcal{C}(\epsilon - \epsilon_0).$$

4 Ferromagnetism

Materials can exhibit different kinds of magnetic ordering, among them para-, ferro- and anti-ferromagnetism (see for example [44, 45, 46]), that refer to the ordering of so called *magnetic moments* in the material: These can be totally unordered, or aligned parallel or anti-parallel to their neighboring moments. In this work, *ferromagnetic materials* are considered. These show, below the critical Curie temperature T_{Curie}, a spontaneous long-range ordering of the magnetic moments. This long-range order occurs even if no external field is present. This leads to the spontaneous formation of *magnetic domains*[1], that are magnetic regions of parallel oriented moments, separated by *domain walls* of definite width, where the magnetization gradually changes. The thermodynamics of magnetic processes and magnetism are discussed in the books of Callen [47], Plischke et al. [48] or O'Handley [46], the theory of magnetic domains is explained in the book of Hubert and Schäfer [44]. The following sections introduce and discuss the theory of micromagnetics and the free energies needed to describe the evolution of magnetic moments to the extend needed in this work.

4.1 Constitutive relations and Maxwell's equations

The constitutive relations of magnetism describe how a material responds to changes in a magnetic or electric field. This section follows the book of O'Handley [46]. SI units[2] are used for the units of the physical quantities.

---

[1]or Weiß domains

[2]The International System of Units (from the French Système international d'unités) using meters, kilograms and seconds as basic physical units. According to

The most important quantities in the description of magnetic phenomena are the electric field $\mathbf{E}$ in volts per meter, the magnetic flux density $\mathbf{B}$ in Tesla, the magnetic field $\mathbf{H}$ in amperes per meter and the magnetic dipole density (or magnetization) $\mathbf{M}$ in amperes per meter. Magnetization $\mathbf{M}$ and field $\mathbf{H}$ are related via the *magnetic susceptibility* $\mathcal{X}_m$ as

$$\mathbf{M} = \mathcal{X}_m \mathbf{H},$$

and the magnetic flux density $\mathbf{B}$ relates to $\mathbf{M}$ and $\mathbf{H}$ via the permeability in the vacuum $\mu_0 = 4\pi \cdot 10^{-7} \frac{\text{Henry}}{m}$ (which is a fundamental constant) by

$$\mathbf{B} = \mu_0(\mathbf{H} + \mathbf{M}) = \mu_0(\mathbf{H} + \mathcal{X}_m \mathbf{H}) = \mu_0(1 + \mathcal{X}_m)\mathbf{H}. \qquad (4.1)$$

The relations between $\mathbf{B}$, $\mathbf{H}$ and $\mathbf{E}$ are given by the famous *Maxwell equations* (see [46]):

$$\nabla \cdot \mathbf{E} = \frac{\rho}{\epsilon} \qquad\qquad \nabla \cdot \mathbf{B} = 0 \qquad (4.2)$$

$$\nabla \times \mathbf{E} = -\frac{\partial \mathbf{B}}{\partial t} \qquad\qquad \nabla \times \mathbf{B} = \mu_0 \mathbf{J} + \frac{\mu_0 \epsilon \partial \mathbf{E}}{\partial t}$$

ρ is the *electric charge density* and ϵ the *vacuum permittivity*.

Assuming magnetostatic and electrostatic situations, the relations simplify because of $\frac{\partial \mathbf{E}}{\partial t} = 0$ and $\frac{\partial \mathbf{B}}{\partial t} = 0$. Fig. 4.1 (taken from the book of Stöhr and Siegmann [50]) shows a diversification of the different interacting fields $\mathbf{B}$, $\mathbf{H}$ and $\mathbf{M}$. The magnetic field $\mathbf{H}$ has contributions from an applied external field $\mathbf{H}_{\text{ext}}$ and the demagnetization field $\mathbf{H}_{\text{demag}}$ that acts inside the body. $\mathbf{H}_{\text{demag}}$ is curl-free (see [44]), i.e. $\nabla \times \mathbf{H}_{\text{demag}} = \mathbf{0}$. When no external magnetic field is present, then $\mathbf{H} = \mathbf{H}_{\text{demag}}$, and from the Helmholtz decomposition theorem Th. 2.6 follows the existence of a scalar potential $\psi : \mathbb{R}^3 \to \mathbb{R}$ such that

$$\mathbf{H}_{\text{demag}} = -\nabla \psi.$$

---

the book of O'Handley [46], using SI units the field $\mathbf{B}$ is considered the most important quantity, in opposition to the cgs unit system, where the magnetization $\mathbf{M}$ is considered more important. For an interesting discussion about different units used in micromagnetics and their interrelation see the article by Scholten [49].

With Eqs. (4.2) and (4.1) follows

$$0 = \nabla \cdot \mathbf{B} = \nabla \cdot \mu_0(\mathbf{M} + \mathbf{H}_{\mathrm{demag}}) = \nabla \cdot \mu_0(\mathbf{M} - \nabla\psi),$$

and with that

$$\nabla \cdot \mathbf{M} = \Delta\psi. \qquad (4.3)$$

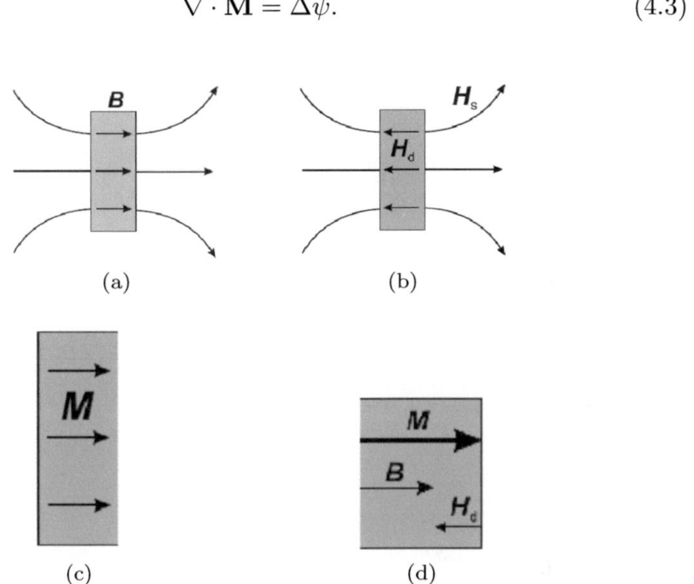

Figure 4.1: The three magnetic fields that act inside and outside a ferro-magnetic body: **(a)** The magnetic flux density $\mathbf{B}$, **(b)** the stray field $\mathbf{H}_s$ outside and demagnetization field $\mathbf{H}_d = \mathbf{H}_{\mathrm{demag}}$ inside a magnetic body, and **(c)** the magnetization $\mathbf{M}$. In **(d)**, the interplay of all fields inside the body is shown. The illustrations are taken from [50].

A general theory of *micromagnetics*, that is the mathematical continuum description of the energies of a magnetic body, goes back to the works of Brown (see [51]). For the following discussion, some general notations and assumptions are fixed. Let $\Omega \subset \mathbb{R}^3$ be a region that will be interpreted as a ferromagnetic body. The magnetization is described by the time and space dependent vector field of spontaneous magnetization $\mathbf{M} : \Omega \to \mathbb{R}^3$.

If isothermal conditions are assumed, the length $|\mathbf{M}|$ of the magnetization vector $\mathbf{M}$ does not change (see [45, 44]) in Ω, such that $|\mathbf{M}(x,t)| \equiv M_S \in \mathbb{R}_{>0}$ for all $x \in \Omega$ and $t \in \mathbb{R}_{\geq 0}$. The scalar value $M_S \in \mathbb{R}^3$ is called the *saturation magnetization*. With that, considerations will be restricted to a unit vector field $\mathbf{m} \equiv \frac{1}{M_S}\mathbf{M}$, and the spontaneous magnetization

$$\mathbf{m} : \Omega \times \mathbb{R}_{\geq 0} \to \mathbb{S}^2, (x,t) \mapsto \mathbf{m}(x,t) = (m_1(x,t), m_2(x,t), m_3(x,t))^T$$

becomes the state variable for the magnetization.

4.2 Ferromagnetic free energy

The free energy density used to describe ferromagnetic effects consists of five micromagnetic contributions: The Zeeman (or external) energy, the demagnetization energy, the exchange energy, the magnetocrystalline anisotropy energy and the magnetostrictive energy, respectively:

$$\begin{aligned} f_{\text{magnetic}}(\mathbf{u}, \mathbf{m}, \nabla\mathbf{m}) = &f_{\text{ext}}(\mathbf{m}) + f_{\text{demag}}(\mathbf{m})+ \qquad\qquad (4.4)\\ &f_{\text{exch}}(\nabla\mathbf{m}) + f_{\text{aniso}}(\mathbf{m})+\\ &f_{\text{m-el}}(\mathbf{u}, \mathbf{m}). \end{aligned}$$

The first two energy densities in Eq. (4.4) are related to magnetostatic effects. The magnetic energy is

$$E_{\text{magnetic}} = \int_{\Omega} f_{\text{magnetic}}\, d\Omega.$$

The different energy density contributions will briefly be discussed now. More detailed explanations can be found e.g. in [45] or [44].

Zeeman energy   The magnetostatic *Zeeman energy* density describes the interaction of the local magnetization $\mathbf{m}$ with an applied external magnetic field $\mathbf{H}_{\text{ext}}$:

$$f_{\text{ext}}(\mathbf{m}) = -\mu_0 M_S(\mathbf{H}_{\text{ext}} \cdot \mathbf{m}). \qquad\qquad (4.5)$$

As can be seen, this energy density is minimized if the magnetic moments align in parallel with $\mathbf{H}_{\text{ext}}$.

Demagnetization energy   The *demagnetization energy* density (or *magnetostatic self energy* density) accounts for the long-range ordering of the magnetic moments. This is reflected by the demagnetization field $\mathbf{H}_{\text{demag}}$ that accounts for the interaction between all local magnetic moments in the system:

$$f_{\text{demag}}(\mathbf{m}) = -\frac{1}{2}\mu_0 M_S(\mathbf{H}_{\text{demag}} \cdot \mathbf{m}).$$

The demagnetization field $\mathbf{H}_{\text{demag}}$ is derived from Maxwell's equations Eqs. (4.2). A solution for $\mathbf{H}_{\text{demag}} = -\nabla\psi$ can be derived in analogy to potential theory in classical electrodynamics (see [19]) from Eq. (4.3). The scalar potential $\psi : \mathbb{R}^3 \to \mathbb{R}$ defined as

$$\psi(\mathbf{r}) = -M_S \int_\Omega \frac{1}{|\mathbf{r} - \mathbf{r}'|}\nabla \cdot \mathbf{m}(\mathbf{r}')\,\mathrm{d}^3\mathbf{r}' + M_S \int_{\partial\Omega} \frac{1}{|\mathbf{r} - \mathbf{r}'|}\mathbf{n}(\mathbf{r}') \cdot \mathbf{m}(\mathbf{r}')\,\mathrm{d}^2\mathbf{r}'$$

is a solution for the scalar potential (cp. [19]). It consists of contributions from the inside of the region Ω and its surface $\partial\Omega$. The field $\mathbf{H}_{\text{demag}}$ can be written explicitly (see [52]) as

$$\begin{aligned}
\mathbf{H}_{\text{demag}}(\mathbf{r}) = &-\frac{1}{4\pi\mu_0}M_S \int_\Omega \nabla \cdot \mathbf{m}(\mathbf{r}')\frac{\mathbf{r} - \mathbf{r}'}{|\mathbf{r} - \mathbf{r}'|^3}\,\mathrm{d}^3\mathbf{r}' \\
&+ \frac{1}{4\pi\mu_0}M_S \int_{\partial\Omega} \mathbf{n}(\mathbf{r}') \cdot \mathbf{m}(\mathbf{r}')\frac{\mathbf{r} - \mathbf{r}'}{|\mathbf{r} - \mathbf{r}'|^3}\,\mathrm{d}^2\mathbf{r}'.
\end{aligned}$$

$\mathbf{n}$ is a vector normal to $\partial\Omega$ pointing outwards. In the case of infinitely periodically extended crystals, the surface term in $\mathbf{H}_{\text{demag}}$ vanishes, but the solution stays valid (cp. e.g. [53]). The assumption of this kind of periodicity is applied when the concept of representative volume elements is adopted. The demagnetization field depends on $\mathbf{m}$, and so on all the local states of the magnetization. This makes its calculation computationally very demanding. To gain efficient calculation methods, special assumptions to $\mathbf{m}$ and its discretization are made to make spectral methods applicable (see Chaps. 7 and 8).

Exchange energy   The short-range magnetic dipole interactions are described by the quantum mechanical *exchange energy density*. It is expressed as the gradient square term

$$f_{\text{exch}}(\nabla\mathbf{m}) = A_{\text{exch}}|\nabla\mathbf{m}|^2, \tag{4.6}$$

where A_{exch} is the material-dependent exchange stiffness constant. $f_{\mathrm{exch}}(\nabla\mathbf{m})$ prefers uniform magnetization states. Using relations from vector calculus, Eq. (4.6) can be shown to be equivalent to (cp. [44, 52])

$$f_{\mathrm{exch}}(\nabla\mathbf{m}) = -A_{\mathrm{exch}}\left(\mathbf{m}\cdot\Delta\mathbf{m}\right),$$

which is a more appropriate expression in some contexts (e.g. when the contribution of $f_{\mathrm{exch}}(\nabla\mathbf{m})$ to the effective magnetic field in the micromagnetic evolution equation is derived).

Magnetocrystalline anisotropy energy The *magnetocrystalline anisotropy density* takes the dependence of the local magnetization on directions of preferred magnetization (the so called *easy axes*) into account. Deviations of the magnetization from the magnetically preferred directions is penalized by f_{aniso}. A special case is the one of *uniaxial anisotropy*, where exactly one easy axis is present. The uniaxial anisotropy energy density reads

$$f_{\mathrm{aniso}}(\mathbf{m}) = K_{\mathrm{aniso}}\left(1-(\mathbf{m}\cdot\mathbf{p})^2\right), \tag{4.7}$$

where K_{aniso} is a material-dependent anisotropy constant, and $\mathbf{p}\in\mathbb{S}^2$ is the direction of the easy axis. In [54] gives a general polynomial expression in terms of even exponents of $(\mathbf{m}\cdot\mathbf{p})$ to model anisotropy for other, non-uniaxial crystal systems.

Magnetoelastic energy The coupling of micromagnetics and elasticity is realized by considering magnetostrictive strains in the elastic energy. This is realized by using the notation for stress-free strain (or eigenstrain) contributions (cp. Chap. 3). If linear strains and the validity of Hooke's law of elasticity are assumed, then

$$f_{\mathrm{m\text{-}el}}(\mathbf{u},\mathbf{m}) = \frac{1}{2}\left((\boldsymbol{\epsilon}(\mathbf{u})-\boldsymbol{\epsilon}_0(\mathbf{m}))\cdot\mathcal{C}(\boldsymbol{\epsilon}(\mathbf{u})-\boldsymbol{\epsilon}_0(\mathbf{m}))\right).$$

$\mathcal{C}$ is the fourth order variant dependent elastic property tensor and $\boldsymbol{\epsilon}(\mathbf{u})$ the second order tensor of total strain (cp. Chap. 3), depending on the displacement field $\mathbf{u}$. The eigenstrains $\boldsymbol{\epsilon}_0$ depend on the magnetization $\mathbf{m}$. The general expression for *magnetostriction* is given by (see [55])

$$\boldsymbol{\epsilon}_0(\mathbf{m}) = \mathcal{N}(\mathbf{m}\otimes\mathbf{m}),$$

where $\mathcal{N}$ denotes the fourth order magnetostrictive property tensor.

4.3 The Landau-Lifshitz-Gilbert equation

The time evolution of the spontaneous magnetization $\mathbf{m}$ is described by the well accepted phenomenological *Landau-Lifshitz-Gilbert equation* (see [56, 44, 12]), that reads

$$\frac{\partial \mathbf{m}}{\partial t} = -\frac{\gamma}{(1 + \alpha_G^2)} \left(\mathbf{m} \times \mathbf{H}_{\text{eff}} + \alpha_G \mathbf{m} \times (\mathbf{m} \times \mathbf{H}_{\text{eff}}) \right), \qquad (4.8)$$

where $\mathbf{H}_{\text{eff}} = \mathbf{H}_{\text{eff}}(\mathbf{m})$ is the *effective magnetic field* that depends on the magnetization $\mathbf{m}$. The effective magnetic field $\mathbf{H}_{\text{eff}}$ arises from energy minimization principles using variational calculus:

$$\mathbf{H}_{\text{eff}} = -\frac{1}{\mu_0 M_s} \frac{\delta E_{\text{magnetic}}}{\delta \mathbf{m}}. \qquad (4.9)$$

E_{magnetic} is the micromagnetic free energy as introduced in the last section. The parameter α_G is the dimensionless phenomenological Gilbert damping constant, γ the gyromagnetic ratio with SI-units $\frac{As}{kg}$. The Eq. (4.8) consists of two parts: The first addend in the brackets describes a gyration of the magnetization $\mathbf{m}$ around the axis given by $\mathbf{H}_{\text{eff}}$, the second addend is a *dissipative Larmor term* that moves $\mathbf{m}$ towards $\mathbf{H}_{\text{eff}}$. Fig. 4.2 shows an illustration for a fixed single magnetic moment. The equation Eq. (4.8) shows that the equilibrium condition

$$\frac{\partial \mathbf{m}}{\partial t} = 0$$

is fulfilled if the magnetic moments are aligned in parallel with the effective field. Using Eq. (4.9) and the definition of the free energies contributing to the magnetic free energy (see Sec. 4.2), the effective magnetic field takes the explicit form

$$\mathbf{H}_{\text{eff}} = \mathbf{H}_{\text{ext}} + \mathbf{H}_{\text{demag}} + \mathbf{H}_{\text{exch}} + \mathbf{H}_{\text{aniso}} + \mathbf{H}_{\text{m-el}}, \qquad (4.10)$$

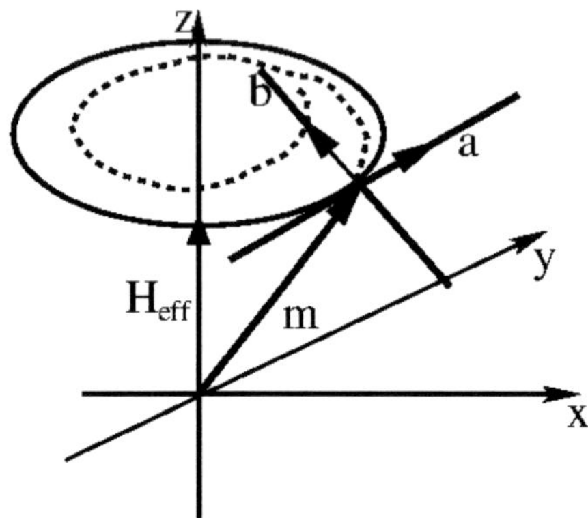

Figure 4.2: Illustration of the Landau-Lifshitz-Gilbert equation for a single magnetization vector **m**, where the effective magnetic field points in the direction of the **z**-axis. **a** represents the precession term, **b** the phenomenological Larmor damping term. In the equilibrium state, the directions of **m** and **H**$_{\text{eff}}$ coincide. This sketch follows [56].

where the addends are gained from the magnetic free energy contributions by variation of E_{magnetic} with **m**. From Eq. (4.6) follows

$$\mathbf{H}_{\text{exch}} = -\frac{1}{\mu_0 M_s}\frac{\partial f_{\text{exch}}(\nabla\mathbf{m})}{\partial\mathbf{m}} = \frac{2A_{\text{exch}}}{\mu_0 M_s}\Delta\mathbf{m}, \qquad (4.11)$$

and from Eq. (4.7)

$$\mathbf{H}_{\text{aniso}} = -\frac{1}{\mu_0 M_s}\frac{\partial f_{\text{aniso}}(\mathbf{m})}{\partial\mathbf{m}} = \frac{2K_{\text{aniso}}}{\mu_0 M_s}(\mathbf{m}\cdot\mathbf{p})\mathbf{p}.$$

The magnetostrictive contribution $\mathbf{H}_{\text{m-el}} = -\frac{1}{\mu_0 M_s}\frac{\delta E_{\text{m-el}}}{\delta\mathbf{m}}$ is not given explicitly, because it can be neglected in the context of magnetic shape memory alloys that are simulated in this work, as the effect of magnetostriction is small compared to the strains arising from the magnetic shape memory effect (cp. [57] or [58]).

The solution of the Landau-Lifshitz-Gilbert equation Eq. (4.8) is, due to the geometric constraint $\mathbf{m} \equiv 1$, a non-trivial task. Problems with ordinary numerical integration schemes will be discussed in Chap. 8, and an adequate numerical integration scheme that is based on the Lie-group theory discussed in Chap. 2 will be introduced.

5 Magnetic shape memory alloys

Magnetic shape memory alloys (MSMAs) are a relatively new class of active and smart materials. They have gained major scientific interest in the last 15 years, since the discovery of the magnetic shape memory effect (MSME) by Ullakko et al. in 1996 [59] in the MSMA Ni_2MnGa. MSMAs allow for giant macroscopic changes in the length of a material induced by an external applied magnetic field, offer a superelastic/superplastic effect and come along with fast actuation times. Their features make them interesting materials from a scientific as well as from an industrial point of view. MSMAs, among which the Heusler alloy Ni_2MnGa is maybe the most famous one, were investigated by different scientific groups on different length and time scales following different objectives. Entel et al. give a good review [60] on the properties of MSMAs. The MSME is a very complex process and incorporates the interplay of elastic and micromagnetic mechanisms on the microscale.

Preceding the MSME is a martensitic transformation, that is a displacive solid-to-solid first order phase transition from a so called austenite phase to a so called martensite phase, assumed to come along with the loss of crystallographic symmetry. The martensitic transformation leads to a martensitic, twin related microstructure of several equivalent martensitic variants. The formal description, classification of twin variants and modeling issues of the martensitic transformation and martensitic microstructures have been analyzed and summarized by Bhattacharya in [10]. Solid-to-solid phase transitions, especially the martensitic transformation and the motion of twin boundaries were investigated by Roytburd and Slutsker [61, 62, 63]. Existing phase-field approaches to model the martensitic transformation include the works of Wang and Khachaturyan [64], Levitas et al. [65] and Kundin et al. [66]

Two continuum theories are essential to describe the MSME properly: The part of continuum mechanics called *kinematics* (see Chap. 3), and the continuum theory called *micromagnetics*, that goes back to the works of Brown [51], especially the field of *ferromagnetism* (see Chap. 4). Continuum mechanics enters into the problem because the material under investigation is completely in the martensitic state when exhibiting the MSME (cp. [67]), and martensitic variants are characterized by their eigenstrains, i.e. the deformation from the parent austenite crystal lattice.

To simulate changes in the magnetic ordering during cooling processes, as occurring during the martensitic transformation when the material transforms from the higher temperature austenite to the lower temperature martensite, the so called Landau-Lifshitz-Bloch equation can be applied. In opposition to Eq. (4.8), the Landau-Lifshitz-Bloch equation accounts for a temperature dependence in the magnetization (see e.g. Garanin [68] or Schieback et al. [69]), and includes thermal fluctuations of the spontaneous magnetization.

In this chapter the most important prerequisites in the modeling of the magnetic shape memory effect are introduced. This includes the martensitic transformation and a brief review of the basic principle of the MSME, i.e. the rearrangement of a martensitic microstructure induced by the application of an external magnetic field. The descriptions of the developing microstructures make use of the continuum theories of the previous chapters.

5.1 The martensitic transformation

The *martensitic transformation* (MT) is a first order diffusionless, displacive, shear-like and reversible phase transition that occurs during cooling. It starts from a high temperature austenite parent phase and results in a lower temperature martensite product phase (see e.g. [70]). The physical parameters determining the MT are four critical temperatures (see [71]).

Martensite start temperature Upon cooling, the temperature T_{ms} where the material contains 1% martensite

Martensite finish temperature Upon cooling, the temperature T_{mf} where the material contains 99% martensite

Austenite start temperature Upon heating, the temperature T_{as} where the material contains 1% austenite

Austenite finish temperature Upon heating, the temperature T_{af} where the material contains 99% austenite

Usually the MT is assumed to be symmetry-breaking (see [37]). That means that the point group of the martensite $\mathcal{P}_m$ is a proper subgroup of the point group $\mathcal{P}_a$ of the austenite: $\mathcal{P}_m < \mathcal{P}_a$. Thus, the MT describes a loss of crystallographic symmetry, as every symmetry operation of the martensite is already contained in the symmetry group of the austenite. This limits the number of allowed transformations (see Fig. 2.3). Starting in the high temperature austenite phase, upon cooling the Bravais lattice of the austenite deforms into the Bravais lattice of the martensite. The material is then, in relation to the parent phase, in a strained state. The deformations are called *Bain strains* and are described by positive definite stretch matrices $U \in \mathbb{R}^{3\times 3}$, where $U = \mathbf{I}$ defines the unstrained austenite state as a reference (cp. [10]). The possible martensitic variants are related to each other via conjugation by elements of $\mathcal{P}_a$: If U is the deformation of one fixed variant, then $M = \{RUR^T | R \in \mathcal{P}_a\}$ is the set of all possible variants (cp. [10] and Fig. 5.1). As $\mathcal{P}_m$ is the symmetry group of the martensite variant described by U, for all $R \in \mathcal{P}_m$ the relation $RUR^T = U$ holds. Using the notation from Chap. 2, $\mathcal{P}_a$ acts on M via conjugation, and M is the orbit of U. Further, U is stabilized by $\mathcal{P}_m$. From Thm. 2 2 follows

$$|M| = |O_U| = \frac{|\mathcal{P}_a|}{|G_U|} = \frac{|\mathcal{P}_a|}{|\mathcal{P}_m|},$$

so the number of possible variants is given by the symmetry-breaking of the transition from the parent phase to the product phase as $\frac{|\mathcal{P}_a|}{|\mathcal{P}_m|}$.

All martensitic variants are energetically equivalent (see Sec. 3.2), what follows directly from material symmetry as variants are related by conjugation. This results in a well-defined energy landscape with energy wells at the austenite and martensite deformation: As a rigid rotation does only change the position of an observer, the austenite state does not only belong to the identical transformation, but also to all rotations $R \in \mathrm{SO}(3)$. Similar, if the variants are characterized by deformations $U_1, \ldots, U_n$, the

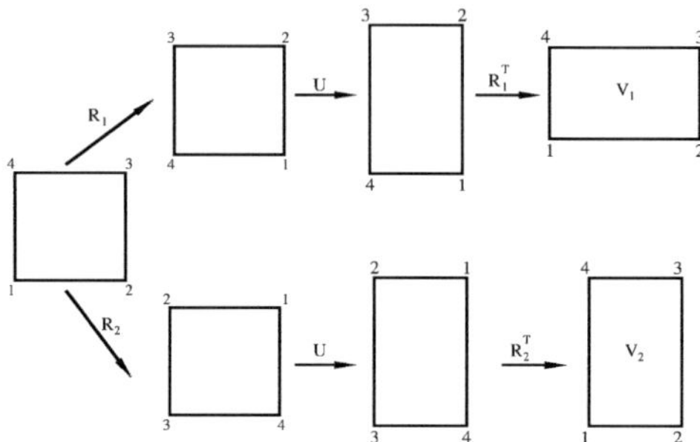

Figure 5.1: Formation of martensitic variants in a simplified 2D cubic-to-tetragonal transformation: The two possible variants are related to each other by conjugation with elements of the 'square' symmetry group, here shown as conjugation of the stretch U with rotations R_1 of $\frac{\pi}{4}$ and R_2 of $\frac{\pi}{2}$. It can easily be seen that variant V_1 also belongs to conjugation with the identity and V_2 to conjugation with a rotation R_3 of $\frac{3\pi}{4}$, so V_1 and V_2 are the only variants.

i-th variant corresponds to all deformations RU_i $(i = 1, \ldots, n)$. This motivates the following definition of *energy wells* (cp. [10]):

Definition 5.1 (ENERGY WELLS) Let $U_1, \ldots, U_n \in \mathbb{R}^{3 \times 3}$ be positive definite stretch deformations characterizing n martensitic variants. Define

$$\mathcal{A} = \mathrm{SO}(3)$$
$$\mathcal{M}_1 = \mathrm{SO}(3)U_1$$
$$\vdots$$
$$\mathcal{M}_n = \mathrm{SO}(3)U_n$$

$\mathcal{A}$ is the *austenite well*, and $\mathcal{M}_i$ $(i = 1, \ldots, n)$ the i-th *martensite well*. □

Clearly, all elements of the same well have the same energy, as frame-indifference states. Furthermore, each martensitic variant lies in exactly one well (see [10]):

Theorem 5.1 (Disjointedness of Energy Wells) Let $U_1, \ldots, U_n \in \mathbb{R}^{3\times3}$ be positive definite stretch deformations characterizing n martensitic variants. Then, each variant belongs to exactly one martensite well, and no variant lies in the austenite well.

Proof. Fix a variant U_i and a rotation $R \in \mathrm{SO}(3)$. Assumption: $RU_i \in \mathcal{M}_j$ for a $j \neq i$. Then there is a rotation $Q \in \mathrm{SO}(3)$ such that $RU_i = QU_j$, or $U_i = R^T Q U_j$. From the polar decomposition theorem Th. 2.5 and $\det U_i > 0$ follows that $R = Q$ and $U_i = U_j$, what contradicts the assumption $i \neq j$. As $U_i \notin \mathrm{SO}(3)$, no martensitic variant can lie in the austenite well. □

The difference between frame-indifference and material symmetry is that the latter acts on the austenite lattice, while the first is applied on the deformed state (see Sec. 3.2).

When the material is at a temperature below the martensitic start temperature T_{ms}, the martensite wells are energetically lower than those of the austenite. Fig. 5.2 shows an illustration of this situation. The orien-

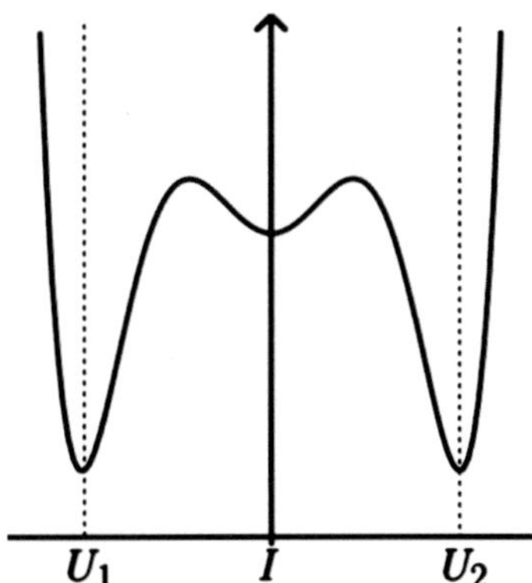

Figure 5.2: The energy landscape of a material at a temperature $T < T_{\mathrm{ms}}$ as a function of deformation from the austenite parent phase: Equivalent global minima exist for two possible martensitic variants. The figure is taken from [5].

tation relation between the martensitic variants is not arbitrary, but also well defined, as the variants determined by U_i and U_j develop from the same parent phase. These deformations are compatible if they obey the rank one kinematic compatibility condition or Hadamard jump condition (cp. [36] and Eq. (3.3))

$$U_i - U_j = (a \otimes n). \tag{5.1}$$

or, in the geometric linear case (cp. Eq. (3.10)),

$$U_i - U_j = \frac{1}{2} (a \otimes n + n \otimes a). \tag{5.2}$$

Then $n \in \mathbb{S}^2$ is the normal to the plane separating the variants i and j, and $a \in \mathbb{R}^3$ indicates direction and magnitude of a simple shear. From Eqs. (5.1) and (5.2) one obtains that the variants are twin related, and the possible twinning modes and directions can be derived from the linear theory presented in [10].

Definition 5.2 (CHARACTERIZATION OF TWINS) Let F and G be two deformations. F and G are called *compatible*, if Eq. (5.1) (or Eq. (5.2) in the geometric linear case) has a solution for $a \in \mathbb{R}^3$ and $n \in \mathbb{S}^2$. Otherwise, the deformations are *incompatible*. Compatible twins are of

Type I if the plane described by $n \in \mathbb{S}^2$ is rational (i.e. the plane includes all lattice points of a definite lattice plane (hkl), cp. Sec. 2.3)

Type II if the shear movement indicated by $a \in \mathbb{R}^3$ is rational (i.e. includes all lattice points $[uvw]$ of a certain direction in the lattice, cp. Sec. 2.3)

and they are

Compound if the twin is of Type I and Type II

□

The Eqs. (5.1) and (5.2) are therefore called the *twinning equation*. As an example the cubic-to-tetragonal MT that occurs in the MSMA Ni_2MnGa in the modulated 5M state is discussed:

Example 5.1 (MARTENSITIC TRANSFORMATION IN $\mathrm{Ni_2MnGa}$) The Heusler alloy $\mathrm{Ni_2MnGa}$ belongs to the space group $\mathrm{F}m\bar{3}m$, and the point group of the cubic parent phase is 432, and has 24 rotation symmetries, while in the point group of the tetragonal martensite product phase only eight rotations are left (cp. Ex. 2.2), so $\frac{24}{8} = 3$ different variants are possible (see Fig. 5.3). The occurring Bain strain matrices are

$$\mathbf{U}_1 = \begin{pmatrix} \beta & 0 & 0 \\ 0 & \alpha & 0 \\ 0 & 0 & \alpha \end{pmatrix}, \quad \mathbf{U}_2 = \begin{pmatrix} \alpha & 0 & 0 \\ 0 & \beta & 0 \\ 0 & 0 & \alpha \end{pmatrix} \tag{5.3}$$

$$\mathbf{U}_3 = \begin{pmatrix} \alpha & 0 & 0 \\ 0 & \alpha & 0 \\ 0 & 0 & \beta \end{pmatrix},$$

where α and β are related to the change of the crystal axes during the MT. The twinning takes place along the $(110)_c$ directions (referred to in the cubic system). All three possible pairs of variants can form a twin boundary (i.e. for all $i, j \in \{1, 2, 3\}$ exists a solution to Eq. (5.1)). The angle of rotation between the shorted c-axes of two variants in $\mathrm{Ni_2MnGa}$ is about $86.5°$ (cp. [72]), and thus can be approximated by $90°$.
All occurring variants are compound. For $i = 1$ and $j = 2$

$$a = \sqrt{2}\frac{\beta^2 - \alpha^2}{\beta^2 + \alpha^2}(-\beta, \alpha, 0)^T \text{ and } n = \frac{1}{\sqrt{2}}(1, 1, 0)^T$$

is a solution for U_1 and U_2 for the twinning equation Eq. (5.1). The invariant plane is of (110) type with respect to the cubic axes system. Analogously, one can find solutions for the other two pairs of martensitic variants (cp.[10]).
Similar results hold for the case of the geometrically linear theory of elasticity. Considering again the variants for $i = 1$ and $j = 2$, solutions for the compatibility equation Eq. (5.2) are found by setting

$$a = \sqrt{2}(\alpha - \beta)(-1, 1, 0)^T \text{ and } n = \frac{1}{\sqrt{2}}(1, 1, 0)^T.$$

Contrary to interfaces between two variants of martensite, interfaces between martensite and austenite are never compatible (but in the unlikely

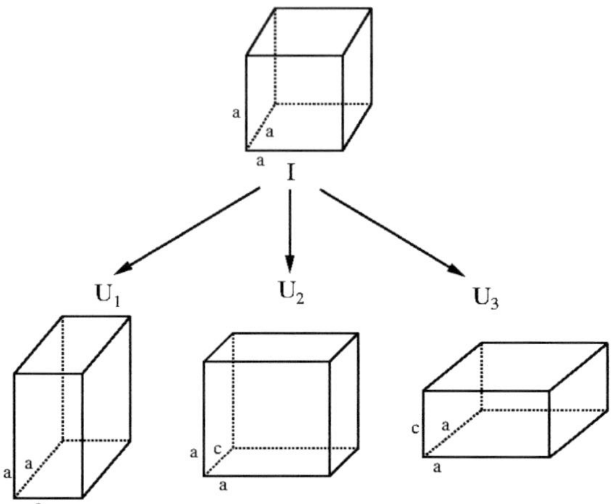

Figure 5.3: The cubic-to-tetragonal MT: Transformation from a cubic austenite parent phase to a tetragonal product phase by changing the cubic axes. Three crystallographically and energetically equivalent variants are possible, described by the Bain matrices U_1, U_2 and U_3. The 3×3 identity matrix $\mathbf{I}$ represents the austenite state.

case of $\beta = 0$ (cp. [35])), i.e. no pair $(a, n) \in \mathbb{R}^3 \times \mathbb{S}^2$ satisfies either the twinning equations

$$\mathbf{I} - U_i = a \otimes n,$$

or

$$U_i - \mathbf{0} = \frac{1}{2} \left(a \otimes n + n \otimes a \right).$$

$\square$

5.2 Conventional and magnetic shape memory effect

The martensitic transformation described in the last section is reversible as, upon heating, a material in the martensitic state transforms back into the austenite state. As there is only one 'austenite variant' the material 'remembers' its former austenite state and with this its shape before the transformation. This effect is called the (conventional or thermal) *shape memory effect* that is the basis of many applications in the automotive industry or medicine (cp. [70, 73]). The shape memory effect comes along with the so called *superelastic effect* (or *pseudoplastic effect*) (see Fig. 5.4a), when the material is in the martensitic state: The ideal material at first behaves linear elastic up to a stress threshold σ_V where the twin boundary motion is induced. This results in a stress plateau. After the material has completely transformed, the material responds again linearly with the modulus of the remaining variant (see [73]).

The *magnetic shape memory effect* is based on the rearrangement a twinned microstructure by externally applying a magnetic field. The MSME takes place completely in the martensitic phase of a ferromagnetic shape memory material (cp. [67]). Magnetic shape memory materials are ferromagnetic hard shape memory materials, and can be used as actuators or dampers that are operated at constant temperature T_{op}. An example is the Heusler alloy Ni_2MnGa that is homogeneous in the sense that the concentration is the same everywhere in the material, and with this the magnetic exchange properties are homogeneous, too. Due to material symmetries the physical properties are the same in equivalent crystallographic directions, and the different martensitic variants are crystallographically and energetically equivalent. By application of an external magnetic field, the Zeeman energy in the material is increased (cp. Eq. (4.5)). As the material is ferromagnetic hard it is energetically more favorable to move the twin boundaries than to move local magnetic moments out of the directions of the easy axes if the external field favors one of the variants (i.e. is aligned with one variants easy axis). This is the basic principle of the MSME, as the systems tends to minimize its energy by aligning local magnetic moments with the direction of the externally applied field. The rearrangement process comes along with giant macroscopic strains as sketched above. The basic functional principle of the MSME is illustrated in Fig. 5.4b.

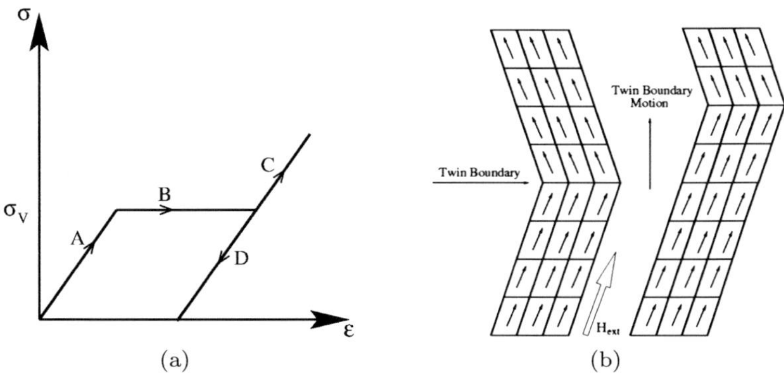

Figure 5.4: **(a)** Pseudo-plastic behavior: A martensitic twinned material exposed to an external mechanical load behaves linear elastic until a threshold σ_V is overcome (**A**) and a phase transition is induced, what results in a stress plateau (**B**), where the strain increases largely at low increasing stress. When the single variant state is reached, the behavior is again linear elastic (**C**), determined by the elastic modulus of the single variant. When no nucleation of variants is induced, the specimen remains strained in the single variant state when the external load is decreased again (**D**).
(b) The basic principle of the motion of twin variants induced by an applied magnetic field in MSME materials in a simplified 2D setting: A material consisting of two martensitic variants is exposed to an external magnetic field applied in direction of the magnetic easy axis of one variant. This increase of Zeeman energy of the other variant together with the high magnetocrystalline anisotropy makes it energetically more favorable to move the twin boundary by transforming the disadvantageous variant into the advantageous one.

Part III

Phase-Field Modeling and Numerical Implementation

6 Phase-field modeling

The phase-field method is a modeling technique to describe the time-spatial evolution of microstructures. In the last decade, it awoke major interest in the field of materials science and has been applied to a variety of different scenarios. In the phase-field method, an artificial order parameter is introduced as an abstract concept to describe the evolution of a microstructure by replacing the sharp interface description of a physical problem by a diffusive interface description in which the phases in the system of interest are separated by interfaces of finite width. The order parameter is coupled to the microscopic or mesoscopic properties of the material. This chapter gives a brief review on the phase-field method in general. The phase-field model published by Nestler et al. in 2005 [1] will be discussed in more detail, as this is the model on which the modeling approach presented in this work is based on.

6.1 Origin of phase-field models

Phase-field models exist for many different applications. They are used to analyze the time-spatial evolution of microstructures on different length scales in different fields, among these spinodal decomposition [74], solidification processes, grain growth and grain coarsening, and solid-to-solid phase transitions like the martensitic transformation [64]. In the literature, several reviews on the phase-field method for special or general purposes can be found, containing many example applications and references to detailed applications and studies. Examples are the articles by L.Q. Chen [75], Qin and Badeshia [76], Moleans et al. [77] or Nestler and Choudhury [78]. In models that describe interfaces between two or more phases (e.g. solid-liquid interfaces in solidification processes) as a sharp transition, compatibility conditions have to be defined and maintained

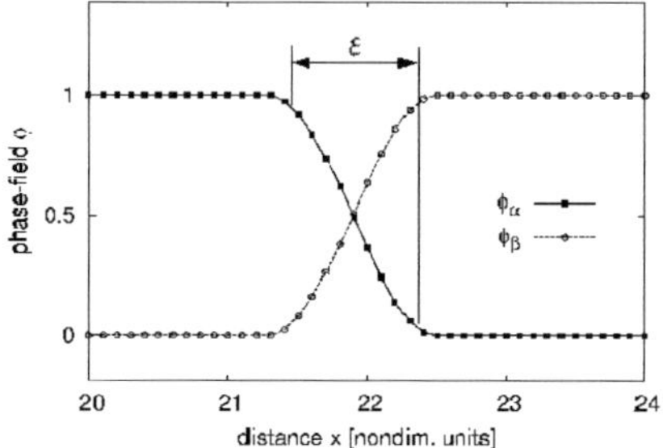

Figure 6.1: Two constraint scalar order parameters ϕ_α and ϕ_β with $\phi_\alpha + \phi_\beta = 1$ varying smoothly from 0 to 1 in an area of a width $\epsilon > 0$. The figure is taken from [3].

during the evolution process. When the boundary is allowed to move in time (one speaks of the *Stefan problem*, cp. [79]), these boundary value problems are hard to solve numerically efficient, because the positions of the interfaces have to be tracked over time to ensure the necessary constraints. This tracking can become a computationally expensive task. All phase-field methods have in common that the sharp interfaces separating different phases in the system under investigation are replaced by *diffusive interfaces*, realized by the introduction of a *diffusive order parameter* or *phase-field parameter* θ (see e.g. [77] for a differentiation of the two concepts) that varies smoothly everywhere in the domain (see Fig. 6.1 for an example). So, θ is a function of space and time. The time-spatial evolution can be described by a set of coupled partial differential equations that can be discretized and solved numerically. The idea of diffusive interfaces goes back to the works of van der Waals in 1893 [80] and Cahn and Hilliard [81]. But also the works on the theory of magnetic domains by Landau and Lifshitz [82] resembles in its main parts the theory of an order parameter (here describing magnetic ordering). Different discretization schemes and optimizations can be applied to implement the phase-field model of choice. In all phase-field methods, the time-evolution of the order parameter θ is derived from variational principles by minimizing the expression for the free energy $\mathcal{F}(\theta, \dots)$ of the system under consideration. This is done by means of the *time-dependent Ginzburg-Landau* or *Allen-Cahn* equation [83, 84]

$$-\frac{\partial \theta}{\partial t} = \tau \frac{\delta \mathcal{F}}{\delta t} = \tau \left(\frac{\partial \mathcal{F}}{\partial \theta} - \frac{\partial \mathcal{F}}{\partial \nabla \theta} \right).$$ (6.1)

The parameter τ is an *interface relaxation parameter* related to the relaxation time of the diffusive interface. The right hand side of Eq. (6.1) results from the Euler-Lagrange formalism of variational calculus. When diffusion processes and the concentrations of K components $c = (c_1, \dots, c_K)$ are considered additionally, the Cahn-Hilliard non-linear diffusion equation

$$\frac{\partial c}{\partial t} = \nabla M \left(\nabla \frac{\delta \mathcal{F}}{\delta c} \right)$$ (6.2)

is additionally solved for the concentration vector c. In Eq. (6.2), M is a matrix related to the interface mobility. Together, both types of equations provide the basic governing equations for phase-field models [84].

6.2 A multi phase-field model with elastic and micromagnetic contributions

A general phase-field model that allows to treat the arising boundary value problems in microstructure modeling has been introduced by Nestler et al. in [1]. This method considers the modeling of the time-spatial evolution of multi-phase multi-component systems that consist of N phases and K components in a region $\Omega \subset \mathbb{R}^3$. A set of non-conserved time and space dependent smooth order parameters with values in the closed interval $[0, 1]$, the so called *phase fields*, is introduced, and collected in the order parameter $\phi = (\phi_1, \dots, \phi_N)^T \in [0, 1]^N$. The *bulk* of a phase $\alpha \in \{1, \dots, N\}$ is defined as the pre-image of one of ϕ_α as $\phi_\alpha^{-1}(1)$. A diffusive interface separates different phases, which is the region where $\alpha \leq N$ exists with $\phi_\alpha \in]0, 1[$. The phase fields locally have to sum up to one (and are interpreted as the local volume fraction of each phase):

$$\sum_{\alpha=1}^{N} \phi_\alpha = 1.$$ (6.3)

The set of allowed states for the order parameter is the *Gibbs simplex*

$$\mathcal{G} = \{\boldsymbol{\phi} = (\phi_1, \ldots, \phi_N)^T \in \mathbb{R}^N \mid \sum_{\alpha=1}^{N} \phi_\alpha = 1 \text{ and} \tag{6.4}$$

$$\phi_\alpha \geq 0 \text{ for all } \alpha \leq N\}.$$

The general integral Helmholtz free energy formulation is of Ginzburg-Landau type and reads

$$\mathcal{F}(\boldsymbol{\phi}, \ldots) = \int_\Omega \left(\xi a(\boldsymbol{\phi}, \nabla\boldsymbol{\phi}) + \frac{1}{\xi} w(\boldsymbol{\phi}) + f(\boldsymbol{\phi}, \ldots) \right) \, d\vec{x}. \tag{6.5}$$

The integral expression depends on all thermodynamic variables of interest via the bulk free energy density term $f(\boldsymbol{\phi}, \ldots)$ (indicated by the dots '...' in the argument list of $\mathcal{F}$ and f). The first two addends in Eq. (6.5) are surface energy contributions that are responsible for the establishment of the diffusive interface of finite width, adjustable via the length parameter $\xi \in \mathbb{R}_{>0}$ (see [85]). The function $a(\boldsymbol{\phi}, \nabla\boldsymbol{\phi})$ is a gradient energy that broadens the interface, while $w(\boldsymbol{\phi})$ is a potential that penalizes pure interfacial states. The potential is non-convex and provides N global minima that correspond to the bulk states of each phase. For $w(\boldsymbol{\phi})$, a higher order variant of a multi-obstacle potential is used that allows to suppress the occurrence of spurious 'third phases' in binary interfaces (see [1]). The bulk free energy $f(\boldsymbol{\phi}, \ldots)$ may depend on several physical quantities, and is defined as the interpolation of individual bulk free energies $f^\alpha(\ldots)$ of each phase α:

$$f(\boldsymbol{\phi}, \ldots) = \sum_{\alpha=1}^{N} h(\phi_\alpha) f^\alpha(\ldots). \tag{6.6}$$

The interpolation function $h : [0, 1] \rightarrow [0, 1]$ has to be continuously differentiable and to satisfy the conditions $h(0) = 0$ and $h(1) = 1$.[1] Valid choices are e.g. $x \mapsto x^2(3 - 2x)$ or $x \mapsto x^3(6x^2 - 15x + 10)$. The evolution equations for the phase fields are based on a modified version of Eq. (6.1) as will be shown below. When the equations of motions are derived from variational methods or material properties are defined, it has to be paid

---

[1] In [84], $\frac{\partial h}{\partial x}(0) = 0 = \frac{\partial h}{\partial x}(1)$ is additionally demanded.

attention to the definition of the total bulk free energy as the interpolation of individual phase bulk free energies, because the interpolation function enters there.

The choice for the *gradient energy function* $a(\phi, \nabla\phi)$ includes the summation over the squared norms of antisymmetric *generalized gradient vectors* $q_{\alpha\beta} = (\phi_\alpha \nabla\phi_\beta - \phi_\beta \nabla\phi_\alpha)$ that span all phase combinations

$$a(\phi, \nabla\phi) = \sum_{\alpha<\beta} \gamma_{\alpha\beta} \left(a_{\alpha\beta}(\phi, \nabla\phi)\right)^2 |q_{\alpha\beta}|^2. \tag{6.7}$$

The gradient vectors $q_{\alpha\beta}$ are well suited to correctly represent the surface energies in multi-phase points (such as triple junctions). Along pure α/β phase-boundaries, $q_{\alpha\beta}$ reduces to $q_{\alpha\beta} = -\nabla\phi_\alpha$. $\gamma_{\alpha\beta}$ in Eq. (6.7) is the surface free energy per unit area (in $\frac{J}{m^2}$) of the α/β boundary, which may additionally depend on the relative orientation of the interface, if appropriate anisotropy functions $a_{\alpha\beta}(\phi, \nabla\phi)$ are given. The choice of $a_{\alpha\beta} \equiv 1$ represents the case of isotropy. Otherwise, $a_{\alpha\beta}(\phi, \nabla\phi)$ represents the anisotropic gamma-plot used in the Wulff construction of the crystal shape. In 2D it is given by $\gamma_{\alpha\beta}a_{\alpha\beta}(\theta)$ for an angle of orientation $\theta \in [0, 2\pi[$. The term has to appear in squared form within the gradient energy Eq. (6.7).[2] When crystal growth from a hydrothermal solution is modeled, it can be desirable to include a strong surface energy anisotropy between dedicated phases. For this case a piece-wise defined function using a maximum condition

$$a_{\alpha\beta}(\phi, \nabla\phi) = \max_{1\leq k\leq n} \left\{ \frac{\vec{q}_{\alpha\beta}}{|\vec{q}_{\alpha\beta}|} \cdot \vec{\eta}_k \right\} \tag{6.8}$$

produces strongly faceted crystals (see [1] and Chap. 9). Here, $\{\vec{\eta}_k | k = 1, \ldots, n\}$ for an $n \in \mathbb{N}$ denotes the complete set of vectors of the corresponding Wulff shape (either 2D or 3D). In Fig. 6.2b a polar plot of this function for a cubic symmetry is given (represented by $n = 6$ and edge vectors $\vec{\eta}_k \in \{\pm e_x, \pm e_y, \pm e_z\}$), together with the evolving octahedral shape of the crystal.

---

[2] The reason for the squared form lies in the calibration property of the interface tension $\bar{\gamma}_{\alpha\beta}$, which can be calculated by integration along a path perpendicular to the interface, leading from phase α to phase β as $\bar{\gamma}_{\alpha\beta} = 2 \int_{-\infty}^{+\infty} \sqrt{a(\phi\nabla\phi)w(\phi)}dx$ [1]. This note is taken from [2].

The chosen potential $w(\phi)$ is the following non-smooth *multi-obstacle potential*:

$$w(\phi) = \begin{cases} \dfrac{16}{\pi^2} \displaystyle\sum_{\substack{\alpha,\beta=1 \\ (\alpha<\beta)}}^{N} \gamma_{\alpha\beta}\, \phi_\alpha \phi_\beta + \displaystyle\sum_{\substack{\alpha,\beta,\gamma=1 \\ (\alpha<\beta<\gamma)}}^{N} \gamma_{\alpha\beta\delta}\, \phi_\alpha \phi_\beta \phi_\delta, & \text{if } \phi \in \mathcal{G}, \\[2em] \infty, & \text{else} \end{cases} \tag{6.9}$$

Each ϕ_α is bound to its definition range by infinite potential walls. The second, higher order term in the first case of Eq. (6.9) includes the summation over all existing $(\alpha/\beta/\delta)$-combinations and modifies the potential to avoid small contributions of other phase fields in a diffuse α/β interface. When curvature driven processes dominate in a simulation, the occurrence of these third phases is adequately suppressed by using a uniform value as $\gamma_{\alpha\beta\delta} = 10 \cdot \max\{\gamma_{\alpha\beta} | \alpha, \beta \leq N\}$ (cp. [86]). The interplay of both surface energy terms, $a(\phi, \nabla\phi)$ and $w(\phi)$, leads to a diffuse interface of definite width, as depicted in Fig. 6.1 for a simulation of a planar α/β front.

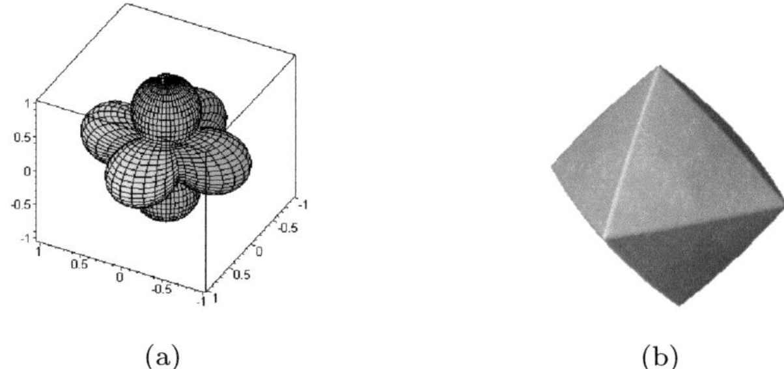

(a)   (b)

Figure 6.2: **(a)** The polar plot of function (6.8) for an anisotropy reflecting cubic symmetry in 3D. **(b)** Under free growth conditions, this anisotropy leads to a crystal with octahedral shape. The figures are taken from [3].

The next subsections specify the free energy contributions to the bulk free energy f as needed to simulate phenomena that include micromagnetic and elastic effects (cp. Chaps. 5 and 10). Additional field parameters are needed to cover the physics of elastic and magnetic processes, namely the elastic displacement field $\mathbf{u}$ and the spontaneous magnetization $\mathbf{m}$ (see

Chaps. 3 and 4). Their evolution equations have to be coupled to the evolution of the phase fields.

6.2.1 Elastic free energy

The elastic free energy f_{elast} is constructed to be the interpolated sum of the elastic energies of the individual phases $\alpha \leq N$. For each phase, the eigenstrain according to Eq. (3.13) has to be considered, giving

$$f_{\text{elast}}^{\alpha} = \frac{1}{2} \left(\epsilon - \epsilon_0^{\alpha} \right) C^{\alpha} \left(\epsilon - \epsilon_0^{\alpha} \right),$$

and consequently

$$f_{\text{elast}} = \sum_{\alpha=1}^{N} h(\phi_\alpha) f_{\text{elast}}^{\alpha} = \sum_{\alpha=1}^{N} h(\phi_\alpha) \frac{1}{2} \left(\epsilon - \epsilon_0^{\alpha} \right) C^{\alpha} \left(\epsilon - \epsilon_0^{\alpha} \right).$$

From this, for the interpolation of the stress follows as

$$\frac{\partial f_{\text{elast}}}{\partial \epsilon} = \sum_{\alpha=1}^{N} h(\phi_\alpha) C^{\alpha} \left(\epsilon - \epsilon_0^{\alpha} \right) = \sum_{\alpha=1}^{N} h(\phi_\alpha) \sigma^{\alpha}$$

with the phase-dependent stresses

$$\sigma^{\alpha} = C^{\alpha} \left(\epsilon - \epsilon_0^{\alpha} \right).$$

6.2.2 Micromagnetic free energy and micromagnetic fields

For the micromagnetic free energy, the same approach as in Eq. (6.6) is taken. The energies introduced in Sec. 4.2 are considered. In the context of this work, only the magnetic anisotropy energy f_{aniso} will depend on variant specific properties, and here only the directions of the easy axes will differ, while the anisotropy constant K_{aniso} will be the same for the different phases. This is justified in the context of Heusler alloys like the magnetic shape memory alloy Ni_2MnGa, as the easy axes of different variants differ, but the exchange properties are the same throughout the material (cp. Chap. 5). If the exchange constant would be different in different phases, additional and more complex boundary conditions for the magnetic exchange would have to be fulfilled (see [44]). Furthermore, magnetostriction will be completely neglected, as the effect is considered to be small compared to the pseudoplastic effects in magnetic shape memory alloys (cp. [87] or [58]). Under this assumptions the micromagnetic energy contributions to the bulk free f energy in the phase-field function reads

$$
f_{\text{magnetic}}(\phi, \mathbf{m}, \nabla \mathbf{m}) = f_{\text{ext}}(\mathbf{m}) + f_{\text{demag}}(\mathbf{m}) +
$$

$$
f_{\text{exch}}(\nabla \mathbf{m}) + \sum_{\alpha=1}^{N} h(\phi_\alpha) f^\alpha_{\text{aniso}}(\mathbf{m})
$$

$$
= - \mu_0 M_S (\mathbf{H}_{\text{ext}} \cdot \mathbf{m}) - \frac{1}{2} \mu_0 M_S (\mathbf{H}_{\text{demag}} \cdot \mathbf{m})
$$

$$
+ A_{\text{exch}} |\nabla \mathbf{m}|^2 + \sum_{\alpha=1}^{N} h(\phi_\alpha) K_{\text{aniso}} \left(1 - (\mathbf{m} \cdot \mathbf{p}^\alpha)^2 \right).
$$

6.2.3 Dynamic equations for the phase-fields

The evolution equations of the phase fields can be derived from the free energy functional $\mathcal{F}$ in Eq. (6.5) by relating the temporal change of the order parameter $\frac{\partial \phi_\alpha}{\partial t} =: \partial_t \phi_\alpha$ to the variational derivative of the functional $\mathcal{F}$, using the Euler-Lagrange formalism (cp. [86] and Eq. (6.1)):

$$
\tau(\phi, \nabla\phi)\, \xi\, \partial_t \phi_\alpha = \xi \left(\nabla \cdot a_{,\nabla\phi_\alpha}(\phi, \nabla\phi) - a_{,\phi_\alpha}(\phi, \nabla\phi) \right) - \tag{6.10}
$$

$$
\frac{1}{\xi} w_{,\phi_\alpha}(\phi) - f_{,\phi_\alpha}(\phi) - \lambda,
$$

with

$$
\lambda = \frac{1}{N} \sum_{\alpha} \left(\xi \left(\nabla \cdot a_{,\nabla\phi_\alpha}(\phi, \nabla\phi) - a_{,\phi_\alpha}(\phi, \nabla\phi) \right) - \tag{6.11}
$$

$$
\frac{1}{\xi} w_{,\phi_\alpha}(\phi) - f_{,\phi_\alpha}(\phi) \right).
$$

The comma-separated subindices on the right hand side of Eq. (6.10) indicate derivations of the function with respect to ϕ_α and the gradient components $\frac{\partial \phi_\alpha}{\partial x_i}$. A kinetic coefficient $\tau(\phi, \nabla\phi)$ is included on the left hand side of the set of reaction-diffusion equations (6.10). This establishes a relationship between growth velocity and driving forces. It is calculated as

$$\tau(\phi, \nabla\phi) = \sum_{\alpha < \beta} g_{\alpha\beta}(\phi)\, \tau_{\alpha\beta}(\phi, \nabla\phi) \tag{6.12}$$

$$\text{with } \tau_{\alpha\beta}(\phi, \nabla\phi) = \tau_{\alpha\beta}^0\, a_{\alpha\beta}^{kin}(\phi, \nabla\phi) \text{ and } g_{\alpha\beta}(\phi) = \frac{\phi_\alpha \phi_\beta}{\sum_{\alpha<\beta} \phi_\alpha \phi_\beta}, \tag{6.13}$$

where $g_{\alpha\beta}(\phi)$ is a normalized interface interpolation function, and $\tau_{\alpha\beta}^0$ is a constant related to the respective phase boundary mobility. Furthermore, the kinetic coefficients $\tau_{\alpha\beta}$ get an orientation dependency by multiplication with an anisotropy function like the one in Eq. (6.8) and other functions to establish kinetic anisotropic behavior.

The coexistence of more than two phase-field parameters imposes additional conditions on their definition range. The value of each ϕ_α is bound to lie in the Gibbs simplex (cp. Eq (6.4)), and the sum constraint Eq. (6.3) has to be ensured. Both constraints, necessary for a correct calibration of energy, are guaranteed by subtracting the Lagrange multiplier Eq. (6.11) on the right hand side of Eq. (6.10). The correctness of this procedure can be seen as follows: Eq. (6.3) implies

$$\sum_{\alpha=1}^{N} \frac{\partial \phi_\alpha}{\partial t} = 0.$$

By inserting the Eqs. (6.10), it can be seen that the constraints are met by means of the Lagrange parameter λ.

6.3 Phase-field methods for solid-state phase transformations

In the literature there exist several phase-field approaches for the modeling of mechanically influenced solid-to-solid phase transformations (see e.g. [88, 53, 89, 58, 90]). Common to these approaches is that they are based on the so called *Landau theory* to reflect the solid-to-solid phase transformation, so this theory will be very briefly sketched in this section. A differentiation of the model presented in Sec. 6.2 from the different modeling approaches based on the Landau theory is given in the next section. The Landau theory assumes a symmetry-breaking phase transition, meaning that the transition starts from a parent phase with symmetry group $\mathcal{P}_\text{parent}$ and ends in a child phase $\mathcal{P}_\text{child}$ with $\mathcal{P}_\text{child} < \mathcal{P}_\text{parent}$ (see [37]). So, every symmetry in the child phase is already present in the parent phase. A new variable Q is introduced to describes the thermodynamics of the system completely. This *order parameter* is an indicator for the symmetry-breaking.[3] Under certain thermodynamic conditions, the phase with lower symmetry might be more stable than the higher symmetric phase.[4] The following explanations and examples strictly follow the book of Salje [37].

The Landau theory is based on a Gibbs-free energy description for the parent and the product phase. The Gibbs-free energy $\mathcal{G}$ depends, in addition to the usual thermodynamic variables like temperature T, pressure P, particle number N etc, on the order parameter Q, so $\mathcal{G} = \mathcal{G}(\ldots, Q)$. The difference between the energies of the high-energy and the low-energy phase is called the *excess Gibbs-free energy*:

$$\mathcal{G}_\text{excess} = \mathcal{G}_\text{parent} - \mathcal{G}_\text{child}. \tag{6.14}$$

The equilibrium condition becomes

$$\frac{\partial \mathcal{G}}{\partial Q} = 0. \tag{6.15}$$

The high-symmetry phase can be defined by $\mathcal{G} = 0$, the trivial solution of Eq. (6.15). Consequently, $\mathcal{G} \neq 0$ in the low symmetry phase. All physical quantities are now measured with respect to this dedicated high symmetry phase and are called *excess quantities*. The idea of Landau and Lifshitz was the expansion of the excess energy Eq. (6.14) analytically

---

[3]In this section the order parameter is used to describe solid-state transformations, so it is denoted by Q and not by the symbol θ that is reserved for general order or phase-field parameters.

[4]Cp. for example the martensitic transformation in Ni_2MnGa, where the martensitic phase is at low temperatures favored over the austenite phase.

as a Taylor series around the order parameter Q (see [91]).[5] The actual form of the resulting *Landau polynomial* is dictated by the symmetries the high-symmetry phase provides [37]. To give a picture, the simplest forms of the Gibbs potentials according to Salje [37] are briefly discussed here, restricted to materials that provide a spontaneous strain ordering.

The excess Gibbs-free energy is written in terms of an excess enthalpy contribution H and an excess entropy contribution S:

$$G = H - TS. \tag{6.16}$$

The excess enthalpy accounts for long-range elastic interactions and has the form

$$H = -\frac{1}{2}AT_C Q^2 + \frac{1}{4}BQ^4 + \frac{1}{6}CQ^6 + PV.$$

The parameter T_C is related to the transition temperature, while A, B and C are coefficients arising from the Taylor expansion. The effect of pressure can be ignored (because the main focus lies on the effect of temperature), so the term PV vanishes.[6] The excess entropy in its simplest form reads

$$S = -\frac{1}{2}AQ^2.$$

So, for the excess Gibbs-free energy follows from Eq. (6.16)

$$G = H - TS = -\frac{1}{2}A(T - T_C)Q^2 + \frac{1}{4}BQ^4 + \frac{1}{6}CQ^6.$$

This expression is called, according to the arising powers of the order parameter, a *2-4-6 potential*.

Other commonly used Gibbs potentials are a *1-2-3 potential*

$$G = -HQ + \frac{1}{2}A(T - T_C)Q^2 + \frac{1}{3}BQ^3$$

or a *symmetry adapted 2-3-4 potential*

$$G = -\frac{1}{2}A(T - T_C)Q^2 + \frac{1}{3}BQ^3 + \frac{1}{4}CQ^4.$$

---

[5]Q does not need to be 'small' in this expansion, so some authors refer to this derivation as *Landau-like expansion*.

[6]In principle, PV is proportional to PQ^2.

The latter is often used for crystals showing cubic, hexagonal or trigonal symmetry.

More detailed explanations about these potentials are given in [37]. As the Gibbs-free energy $\mathcal{G}$ has to fulfill symmetry relations on the microscale, group theoretic approaches can be applies, connected to minimal representations of the symmetry group (see e.g. the book of Bradley and Cracknell [13] or the article by Cracknell [92]).

To cover the thermodynamics of crystals that exhibit spontaneous strain ordering correctly in the above sense, the *coupling theory* has to be applied, which assumes the crystal to be in thermodynamic equilibrium with a surrounding heat bath, and that any structural transition induces a loss of energy (see [37, 93]). The Gibbs free energy $\mathcal{G}$ then consists of three parts: The Landau potential as introduced above (noted $L(Q)$), the elastic energy $f_{\text{elast}} = \frac{1}{2}\sum_{i,j=1}^{3}\frac{1}{2}\epsilon_{ij}\mathcal{C}_{ijkl}\epsilon_{kl}$ as introduced in Chap. 3, and an interaction energy that couples the order parameter Q and the components of the elastic strain tensor ϵ, weighted by *coupling coefficients* ζ_{ij}:

$$\mathcal{G}(Q,\epsilon) = L(Q) + f_{\text{elast}} + \sum_{i,j}\sum_{m,n}(\zeta_{ij})_{mn}\epsilon_{ij}^{m}Q^{n}.$$

The integer range of the indices m and n is dictated by the symmetries that the high symmetry phase provides. The most simple example of coupling is a *bilinear coupling*

$$\mathcal{G}(Q,\epsilon) = L(Q) + f_{\text{elast}} + \sum_{i,j=1}^{3}\zeta_{ij}\epsilon_{ij}Q.$$

Here, only those components ϵ_{ij} of the strain tensor ϵ appear that are allowed by symmetry.

6.4 Phase-field models for the magnetic shape memory effect

The simulations carried out in Chap. 10 of this work deal with phenomena related to the magnetic shape memory effect. The model as presented in

Sec. 6.2 is applied there. As there exist other phase-field approaches to model the behavior of magnetic shape memory alloys, a brief discussion on some phase-field models that are often cited in the literature is given here. Characteristic and crucial for each phase-field model is the choice of the order parameter and the interpolation of the material properties. The discussion here focuses on phase-field models for the magnetic shape memory effect and related phenomena in the material system Ni_2MnGa. The phase-field models published by Zhang and Chen [53, 88], Jin [58] and Wu et al. [90, 94] construct the free energy of the MSMA based on the Landau theory for solid-to-solid phase transformations. Zhang and Chen use the eigenstrains of the martensitic variants (i.e. the deformation from an undeformed austenite state) as an order parameter that characterizes the martensitic variants, and define a Landau polynomial in terms of symmetry-adapted strain components as proposed by Vasil'ev et al. [95] to couple the order parameter and the elastic strain. The coupling of the order parameter and the magnetization is realized by a magnetoelastic coupling term. The work of Jin [58] defines an order parameter reflecting the three different martensitic variants in analogy to Artemev et al. [89], and a fourth-order Landau polynomial is constructed that provides global minima for the order parameter for martensitic variants at the standard base vectors $\mathbf{e}_i \in \mathbb{R}^3$. The model presented by Wu et al. [90, 94] is an extension of the model of Zhang and Chen [53]. Another phase-field approach has been published by Li et al. [96], where an order parameter related to the volume fractions of the martensitic variants is defined. For the sake of simplicity of the model, a potential accounting for fourth orders of the order parameter is used instead of constructing a Landau polynomial. Using this potential, the energy landscape with wells for the martensitic variants in MSMAs is expressed explicitly. The stated models, except for the one of Jin [58], consider the magnetoelastic coupling, Zhang and Chen [53, 88] and Wu et al. [90, 94] in terms of an explicit coupling term, Li et al. as an additional contribution to the transformation strains ϵ_0. In principle, the same consideration could be added to the model formulation presented by Jin [58]. Jin and Li et al. directly interpolate the material properties in terms of the order parameter, while Zhang and Chen formulate a full expression for the free energy of the magnetic shape memory alloy in terms of the order parameter that varies smoothly in the calculation domain (including a Landau polynomial expression). A different approach is published by Landis in [97]. There, a diffusive interface model based on a continuum theory is proposed. The order parameters are the martensitic free-strain and the magnetization. Sets of generalized *micro-forces* and balance equations corresponding to these order parameters are postulated. The micro-forces do work as the order parameters change. Simulations are carried out using the finite element method to analyze the microstructure evolution in ferromagnetic shape memory alloys. The theory proves well to predict blocking stresses in ferromagnetic shape memory alloys. The model used in the work at hand is based on the model presented by Nestler et al. [1] and uses an order parameter directly related to the volume fraction of the martensitic variants. All material properties and governing equations are consequently derived from the interpolation of free energies using an interpolation polynomial $h : [0, 1] \rightarrow [0, 1]$. Magnetostriction is assumed to be small and therefore neglected in this model with the same arguments as given in [58].

7 Numerical implementation and boundary conditions

The phase-field model introduced in Sec. 6.2 is numerically implemented in the software framework called Pace3D. This software is developed and maintained by the group of Prof. Nestler at the Karlsruhe Institute of Technology (KIT) - Institute of Applied Materials and the Karlsruhe University of Applied Sciences - Institute of Materials and Processes (IMP). The numerical implementation is based on finite differences, and for the solution of the equations of motion explicit forward Euler schemes for the time update are used. Depending on the type of equation to be solved and the constraints that have to be fulfilled, the explicit schemes are chosen adequately. The solver software is written in the programming language C and the code is parallelized using the Message Passing Interface (MPI) library, so that the simulations done in this work could be performed on single and multi-processor PCs as well as on a Linux server cluster. The general numerical techniques to solve the equations of motion for the phase fields and the elastic displacement field are briefly discussed in the next sections. The special techniques to perform micromagnetic calculations efficiently is postponed to the next chapter. The implemented solution procedures for the phase fields and the elastodynamic wave equation in its basic parts are part of the implementation of Pace3D (cp. [98, 99, 100]). The solution procedure for the elastic wave equation has been generalized within this work to make the consideration of elastic properties of arbitrarily oriented phases possible (cp. Appendix B).

7.1 General techniques

For general introductions to numerical mathematics see the textbooks of
Schwarz [101] or Stoer [102]. The basic discretization scheme used to discretize all appearing equations, both in space and time, is based on *finite
differences* (see e.g. the book of Stoer [102]). Spatially, a fixed discrete
point grid is assumed for the calculation domain $\Omega \subset \mathbb{R}^3$, as well as a
frame of reference $(\mathbf{0}, \mathbf{x}_1, \mathbf{x}_2, \mathbf{x}_3)$. This special grid serves as a reference
grid and will be referred to as a *collocated grid*. The points of the grid
are indexed by triples $(i, j, k) \in \mathbb{N}^3$. The values $i_{\max}, j_{\max}, k_{\max}$ indicate
upper boundary cells, 0 cells at the lower boundary. Thus, $(i, j, k) \in \mathbb{N}^3$
is 'inside' the calculation domain, if $0 < i < i_{\max}$, $0 < j < j_{\max}$ and
$0 < k < k_{\max}$, and a *boundary cell* otherwise. The distance between
two neighboring grid points is $\Delta x_i > 0$ in the $\mathbf{x}_i$-direction. The same
approach is taken for the time discretization, where the discrete times t_n
and t_{n+1} are separated by the discrete time-step width $\Delta t > 0$. Usually,
all occurring physical quantities q are functions of space and time, i.e.
$q : \Omega \times \mathbb{R}_{>0} \to M, q \mapsto q(x_1, x_2, x_3, t)$, where M is the set of valid values
of for the quantity q (e.g. $M = \mathbb{R}$ or $M = \mathbb{R}^3$ or $M = \mathbb{S}^2$ if q is the temperature, displacement field or magnetization, respectively). Each quantity
is assumed to be given for each discrete time t_n by its values on the grid
points. Occurring spatial derivatives, such as the gradient or the divergence of q (∇q or $\nabla \cdot q$), are approximated by considering the values of q
on these discrete point grid. This gives a restriction on the grid spacings
Δx_i and time spacings Δt to maintain numerical accuracy. Depending on
how many neighboring points are taken into account to approximate the
value of q at a fixed point, the accuracy and stability of the approximation
can be increased. The spatial and time discretization schemes used here
are first or second order accurate, what means the the approximation er-
ror is of magnitude $\mathcal{O}(\max(\Delta x_i)^2)$ or $\mathcal{O}(\max(\Delta t))$.[1] As time integration
schemes, *explicit forward Euler schemes* are used to compute $\dot{q} = \frac{\partial q}{\partial t}$ of
the quantity q (see e.g. the book of Stoer [102]). For the phase fields
ϕ_α and the displacement field components $\mathbf{u}_i$ standard schemes are used.

---

[1] The 'Big-O-Notation' for complexity classes is used rather intuitively here. Meant
is that the error approaches zero with a rate faster or equal to the rate the argument of
$\mathcal{O}$ goes to zero. More formally, $\mathcal{O}$ is the basis of asymptotic measures. The argument of
$\mathcal{O}$ is a function f and $\mathcal{O}(f) = \{g : \mathbb{N} \to \mathbb{N} | \exists n_0 \in \mathbb{N} \; \exists c \in \mathbb{R} \; \forall n \in \mathbb{N}_{>n_0} : g(n) \leq cf(n)\}$
is a class of functions.

The explicit Euler scheme used for the time integration of the spontaneous magnetization $\mathbf{m}$ is, due to the geometric constraint $\mathbf{m} \equiv 1$, based on a geometric integration method and discussed in the next chapter.

7.2 Phase-field equation and boundary conditions for the phase fields

The phase-field equation Eq. (6.10) is discretized using finite differences, and an explicit forward Euler scheme is used to numerically integrate the equations as described by Nestler [98]. The computationally most demanding term of Eq. (6.10), $\nabla \cdot a_{,\nabla\phi_\alpha}(\boldsymbol{\phi}, \nabla\boldsymbol{\phi})$, is split up into the primary calculation of the surface energy flux $\xi a_{,\nabla\phi_\alpha}$ as a function of the generalized gradient vectors $q_{\alpha\beta}$, and a secondary step in which the divergence operation $(\nabla\cdot)$ is calculated. Furthermore, the calculations are limited to the diffuse interface, where non-vanishing gradients can occur. The usual resolution of the diffusive interface should be chosen to be about ten grid points (cp. Chap. 6.2).

The boundary conditions for the phase-fields ϕ_α used in this work either reflect periodicity of the geometry, or are of the special Neumann-type $\frac{\partial\phi_\alpha}{\partial n} = 0$ (where n is the unit normal on the boundary $\partial\Omega$ pointing outwards). The periodic boundary conditions are realized by copying values from the first lower (or upper last) non-boundary layers into the layers of the opposite boundary. The special Neumann boundary condition is realized by copying the last non-boundary layer into the boundary in the direction of n, forcing the gradients of the phase fields to vanish. This boundary condition alters the angles that interfaces enclose with the boundary if interface and boundary normal n are neither parallel nor orthogonal, as it enforces the parallel alignment of phase boundaries.

As a special technique to reduce the computation time, a method to limit locally the temporal update to a small subset of the phase fields of fixed cardinality has been developed by Nestler et al. [103] The algorithm is based on the approach described by Kim et al. [104], where is shown that a selection of the five dominant phase-field variables in 2D and six in 3D per grid point is sufficient and does not significantly reduce the accuracy of the simulation results. Hence, the computation time becomes independent of the number of phase-field parameters, and increases only linearly with the total number of grid points in the simulation domain. This is an important technique when many phases are present in the system as in the simulations carried out in Chap. 9.

7.3 Elastic equation

The displacement field $\mathbf{u} : \Omega \times \mathbb{R}_{\geq 0} \to \mathbb{R}^3$ is a vector valued quantity, so *staggered grids* are used to increase the numerical stability of the integration schemes (see [105] or [106]). A new grid for each component $\mathbf{u}_i$ of the displacement field $\mathbf{u}$ is introduced, shifted by $\frac{1}{2}\Delta x_i$ in the $\mathbf{x}_i$-direction. The resulting interrelation of the four grids can be thought to span rectangular grid cells, where the phase field values lie on the collocated grid in the center of the cell, and the values for the displacement field components lie on the centers of the rectangles' faces (in Fig. 7.1 an illustration is shown). This approach may lead to the necessity of computing the values of phase fields or displacement field values at grid positions that differ from the grid positions on which they are stored. For example, the values of the strain tensor $\boldsymbol{\epsilon} = \frac{1}{2}(\nabla\mathbf{u} + \nabla\mathbf{u}^T)$ are needed on the grid positions where the phase field values are stored, as will be shown below. These computations are done by interpolation procedures that become computationally demanding when these evaluations have to be carried out frequently.

7.3.1 Discretization of the elastic equation

To compute all terms that are related to the solution of the elastic equation, the Voigt notation, that is the representation of the elastic stress, strain and stiffness tensor as 6-vectors and 6×6 matrix are applied (see Appendix B)[2]. The entries of the matrix representing the elastic stiffness

---

[2]For convenience, the notation used here in this section uses the matrix notation for the elastic stiffness, but the doubly indexed tensor notation for the elastic stress and strain.

tensor of phase α are denoted by c_{ij}^{α} with $i, j \in \{1, \ldots, 6\}$. The elastodynamic wave equation for the evolution of the displacement fields is gained from (cp. Eq. (3.13))

$$\rho\ddot{\mathbf{u}} = -\frac{\delta E_{\text{elast}}}{\delta\mathbf{u}} = \nabla \cdot \boldsymbol{\sigma} \tag{7.1}$$

as shown by Spatschek et al. in [99]. Eq. (7.1) is solved for each component of $\mathbf{u}$. To account for the dissipation of elastic energy, a damping term is introduced that is proportional to the velocity of the displacements, $\dot{\mathbf{u}}$:

$$\rho\ddot{\mathbf{u}} + \kappa\dot{\mathbf{u}} = \nabla \cdot \boldsymbol{\sigma}. \tag{7.2}$$

$\kappa \in \mathbb{R}_{\geq 0}$ is the damping constant, and the damping term is used to damp out small wavelength elastic excitations obliterating the simulation results (as motivated in [100]). The right hand side of Eq. (7.2) involves the computation of the divergence of the elastic stress tensor $\boldsymbol{\sigma}$. As a staggered grid is underlying the discretization scheme, the values that are needed to compute the time update have to be on the correct grid for component $\mathbf{u}_i$ ($i = 1, 2, 3$), what makes the interpolation of components necessary. To gain a second order accurate central differences scheme for the computation of the divergence operator on the grids for the displacement components, the components of the stress tensor $\boldsymbol{\sigma}$ are either needed in the center of the cells (for the diagonal components σ_{ii}) or at the centers of the edges of the cells (for the non-diagonal components σ_{ij} ($i \neq j$)). As the validity of Hooke's law is assumed (see Chaps. 3 and 6), the relation $\boldsymbol{\sigma} = \mathcal{C}(\boldsymbol{\epsilon} - \boldsymbol{\epsilon_0})$ is valid, and from the assumption of linear elasticity $\boldsymbol{\epsilon} = \frac{1}{2}(\nabla\mathbf{u} + \nabla\mathbf{u}^T)$ follows $\boldsymbol{\sigma} = \boldsymbol{\sigma}(\mathbf{u})$. Then

$$\rho\ddot{\mathbf{u}}_i + \kappa\dot{\mathbf{u}}_i = (\nabla \cdot \boldsymbol{\sigma}(\mathbf{u}))_{i,\cdot}, \tag{7.3}$$

i.e. the time update of the i-th component of $\mathbf{u}$ depends on the i-th row of $\nabla \cdot \boldsymbol{\sigma}(\mathbf{u})$. A discrete scheme to compute the $(n+1)$-th time step of the displacement by using second order central differences is given by

$$\mathbf{u}^{n+1} = \frac{1}{\rho + \Delta t\kappa} \left((\Delta t)^2 \nabla \cdot \boldsymbol{\sigma} + (2\rho + \Delta t\kappa)\mathbf{u}^n - \rho\mathbf{u}^{n-1}\right). \tag{7.4}$$

The crucial part is the discretization of the divergence $(\nabla \cdot \boldsymbol{\sigma}(\mathbf{u}))_{i,\cdot}$ on the $\mathbf{u}_i$-grid. Keeping in mind the interpolation of free energies in the phase-field model (see Eq. (6.6)) and the definition of the stresses, the expression

$$\boldsymbol{\sigma}(\mathbf{u}) = \frac{\partial f_{\text{elast}}}{\partial \boldsymbol{\epsilon}(\mathbf{u})} = \sum_{\alpha=1}^{N} h(\phi_\alpha) \frac{\partial f_{\text{elast}}^\alpha}{\partial \boldsymbol{\epsilon}(\mathbf{u})} = \sum_{\alpha=1}^{N} h(\phi_\alpha) \boldsymbol{\sigma}^\alpha(\mathbf{u})$$

$$= \sum_{\alpha=1}^{N} h(\phi_\alpha) \mathcal{C}^\alpha \left(\frac{1}{2}(\nabla \mathbf{u} + \nabla \mathbf{u}^T) - \boldsymbol{\epsilon}_0^\alpha \right)$$

shows that the values of the phase fields and of the components of $\mathbf{u}$ have to be computed on positions that do not lie on the quantities' original grid.

An example is shown to make the arguments more clear. To increase the readability, subscripts at quantities will refer to cell indexes, will superscripts index the components. Let $m \in \{1,2,3\}$ and (i,j,k) the index triple for a cell not on the boundary. The divergence of the m-th row of the divergence of $\boldsymbol{\sigma}$ on the grid of $\mathbf{u}_m$ is approximated by central differences as

$$(\nabla \cdot \boldsymbol{\sigma}_{i,j,k})_{m,\cdot} \approx \frac{\boldsymbol{\sigma}_{i,j,k}^{m1} - \boldsymbol{\sigma}_{i-1,j,k}^{m1}}{\Delta x_1} + \tag{7.5}$$

$$\frac{\boldsymbol{\sigma}_{i,j,k}^{m2} - \boldsymbol{\sigma}_{i,j-1,k}^{m2}}{\Delta x_2} +$$

$$\frac{\boldsymbol{\sigma}_{i,j,k}^{m3} - \boldsymbol{\sigma}_{i,j,k-1}^{m3}}{\Delta x_3},$$

When the expression is evaluated, for the components $\boldsymbol{\sigma}_{i,j,k}^{mn}$ the staggered grid structure has to be respected. As stated above, the components $\boldsymbol{\sigma}_{i,j,k}^{mm}$ are to be calculated on the collocated grid for the phase fields, and the components $\boldsymbol{\sigma}_{i,j,k}^{mn}$ with $m \neq n$ on the centers the cell edges. Explicitly, the discretization is shown for the update of $\mathbf{u}_1$ in a the cell (i,j,k) not on the boundary, and component $\boldsymbol{\sigma}_{i,j,k}^{11}$ of the stress tensor. The components of the strain tensor are symmetric in the chosen discretization[3] and needed in the centers of the cells, the components of the stress tensor $\boldsymbol{\sigma}_{i,j,k}^{11}$ in

---

[3]This fact is not self-evident by can be understood by comparing the components ϵ_{ij} and ϵ_{ji} explicitly in their discretized versions.

cell $(i+1, j, k)$. The strain components in the center of the cell $(i+1, j, k)$ read

$$\epsilon_{i,j,k}^{11} = \frac{u_{i+1,j,k}^{1} - u_{i,j,k}^{1}}{\Delta x_1}$$

$$\epsilon_{i,j,k}^{22} = \frac{u_{i+1,j,k}^{2} - u_{i+1,j-1,k}^{2}}{\Delta x_2}$$

$$\epsilon_{i,j,k}^{33} = \frac{u_{i+1,j,k}^{3} - u_{i+1,j,k-1}^{3}}{\Delta x_3}$$

$$\epsilon_{i,j,k}^{23} = \frac{1}{8}\left(\frac{(u_{i+1,j,k+1}^{2} - u_{i+1,j,k-1}^{2}) + (u_{i+1,j-1,k+1}^{2} - u_{i+1,j-1,k-1}^{2})}{\Delta x_3} + \right.$$
$$\left.\frac{(u_{i+1,j+1,k}^{3} - u_{i+1,j-1,k}^{3}) + (u_{i+1,j+1,k-1}^{3} - u_{i,j-1,k-1}^{3})}{\Delta x_2}\right)$$

$$\epsilon_{i,j,k}^{13} = \frac{1}{8}\left(\frac{(u_{i+1,j,k+1}^{1} - u_{i+1,j,k-1}^{1}) + (u_{i,j-1,k+1}^{1} - u_{i,j-1,k-1}^{1})}{\Delta \mathbf{x}_3} + \right.$$
$$\left.\frac{(u_{i+2,j,k}^{3} - u_{i,j,k}^{3}) + (u_{i+2,j,k-1}^{3} - u_{i,j,k-1}^{3})}{\Delta x_1}\right)$$

$$\epsilon_{i,j,k}^{12} = \frac{1}{8}\left(\frac{(u_{i+1,j+1,k}^{1} - u_{i+1,j-1,k}^{1}) + (u_{i,j+1,k}^{1} - u_{i,j-1,k}^{1})}{\Delta \mathbf{x}_2} + \right.$$
$$\left.\frac{(u_{i+2,j,k}^{2} - u_{i,j,k}^{2}) + (u_{i+2,j-1,k}^{3} - u_{i,j-1,k}^{3})}{\Delta x_1}\right)$$

Accounting for the eigenstrain contributions of different phases (which are constant phase dependent properties), $\sigma_{i,j,k}^{11}$ becomes

$$\sigma_{i,j,k}^{11} = \sum_{\alpha=1}^{N}\left(c_{11}^{\alpha}(\epsilon_{i,j,k}^{11} - \epsilon_0^{\alpha,11}) + \right.$$
$$c_{12}^{\alpha}(\epsilon_{i,j,k}^{22} - \epsilon_0^{\alpha,22}) +$$
$$c_{13}^{\alpha}(\epsilon_{i,j,k}^{33} - \epsilon_0^{\alpha,33}) +$$
$$2c_{14}^{\alpha}(\epsilon_{i,j,k}^{23} - \epsilon_0^{\alpha,23}) +$$
$$2c_{15}^{\alpha}(\epsilon_{i,j,k}^{13} - \epsilon_0^{\alpha,13}) +$$
$$\left. 2c_{16}^{\alpha}(\epsilon_{i,j,k}^{12} - \epsilon_0^{\alpha,12})\right) h(\phi_{i+1,j,k}^{\alpha}).$$

The other components of $\mathbf{u}$ and $\boldsymbol{\sigma}$ of Eq. (7.5) are treated analogously.

When the mechanical equilibrium condition Eq. (3.8) is assumed, the same discretization scheme as introduced above is applied. The idea of calculating the mechanical equilibrium is based on the approach described by Hattel and Hansen [107]. The solution scheme starts from Eq. (7.3) by omitting the damping term

$$(\nabla \cdot \boldsymbol{\sigma}(\mathbf{u}))_{m,\cdot} = 0. \tag{7.6}$$

Discretizing Eq. (7.6) using the same scheme as in Eq. (7.5), the m-th component of $\mathbf{u}_{i,j,k}^{m}$ at grid point (i,j,k) can be explicitly calculated. Again, one has to keep in mind the interpolation of stresses arising from the approach of interpolating free energies in the phase-field model (see Eq. (6.6)). To write the explicit formula for the update of the first component $\mathbf{u}_1$, $\mathbf{u}_{i,j,k}^{1}$ is extracted from Eq. (7.6). Fig. 7.1 shows a sketch of the components that enter the update formula for $\mathbf{u}_1$. The following update formula can be gained:

$$u_{i,j,k}^{1} = \frac{1}{d}\left(\frac{S_{i,j,k}^{11} - S_{i-1,j,k}^{11}}{\Delta x_1} + \frac{S_{i,j,k}^{12} - S_{i,j-1,k}^{12}}{\Delta x_2} + \frac{S_{i,j,k}^{13} - S_{i,j,k-1}^{13}}{\Delta x_3}\right).$$

The six addends arise from the discretization of the divergence of the stress tensor $\boldsymbol{\sigma}$, where implicitly a summation over all phase indexes α is assumed (the summation sign is suppressed to increase the readability of the formulae):

$$S_{i,j,k}^{11} = \Bigg[$$

$$\left(c_{11}^{\alpha}\frac{1}{\Delta x_1}u_{i+1,j,k}^{1} - \epsilon_{0}^{11\alpha}\right)$$

$$+ c_{12}^{\alpha}\Big(\frac{1}{\Delta x_2}(u_{i+1,j,k}^{2} - u_{i+1,j-1,k}^{2}) - \epsilon_{0}^{22\alpha}\Big)$$

$$+ c_{13}^{\alpha}\Big(\frac{1}{\Delta x_3}(u_{i+1,j,k}^{3} - u_{i+1,j,k-1}^{3}) - \epsilon_{0}^{33\alpha}\Big)$$

$$+ c_{14}^{\alpha}\Bigg[$$

$$\Big(\frac{1}{4\Delta x_3}(u_{i+1,j,k+1}^{2} - u_{i+1,j,k-1}^{2} + u_{i+1,j-1,k+1}^{2} - u_{i+1,j-1,k-1}^{2})$$

$$+ \frac{1}{4\Delta x_2}(u^3_{i+1,j+1,k} - u^3_{i+1,j-1,k} + u^3_{i+1,j+1,k-1} - u^3_{i+1,j-1,k-1})) - $$

$$2\epsilon^{23\alpha}_0 \Big]$$

$$+ c^\alpha_{15}\Big[\big(\frac{1}{4\Delta x_3}(u^1_{i+1,j,k+1} - u^1_{i+1,j,k-1} + u^1_{i,j,k+1} - u^1_{i,j,k-1})$$

$$+ \frac{1}{4\Delta x_1}(u^3_{i+2,j,k} - u^3_{i,j,k} + u^3_{i+2,j,k-1} - u^3_{i,j,k-1})\big) - 2\epsilon^{13\alpha}_0 \Big]$$

$$+ c^\alpha_{16}\Big[\big(\frac{1}{4\Delta x_2}(u^1_{i+1,j+1,k} - u^1_{i+1,j-1,k} + u^1_{i,j+1,k} - u^1_{i,j-1,k})$$

$$+ \frac{1}{4\Delta x_1}(u^2_{i+2,j,k} - u^2_{i,j,k} + u^2_{i+2,j-1,k} - u^2_{i,j-1,k})\big) - 2\epsilon^{12\alpha}_0 \Big]$$

$$\Big] h(\phi^\alpha_{i+1,j,k}),$$

$$S^{11}_{i-1,j,k} = \Big[$$

$$\big(- c^\alpha_{11}\frac{1}{\Delta x_1}u^1_{i-1,j,k} - \epsilon^{11\alpha}_0\big)$$

$$+ c^\alpha_{12}\big(\frac{1}{\Delta x_2}(u^2_{i,j,k} - u^2_{i,j-1,k}) - \epsilon^{22\alpha}_0\big)$$

$$+ c^\alpha_{13}\big(\frac{1}{\Delta x_3}(u^3_{i,j,k} - u^3_{i,j,k-1}) - \epsilon^{33\alpha}_0\big)$$

$$+ c^\alpha_{14}\Big[\big(\frac{1}{4\Delta x_3}(u^2_{i,j,k+1} - u^2_{i,j,k-1} + u^2_{i,j-1,k+1} - u^2_{i,j-1,k-1})$$

$$+ \frac{1}{4\Delta x_2}(u^3_{i,j+1,k} - u^3_{i,j-1,k} + u^3_{i,j+1,k-1} - u^3_{i,j-1,k-1})\big) - 2\epsilon^{23\alpha}_0 \Big]$$

$$+ c^\alpha_{15}\Big[\big(\frac{1}{4\Delta x_3}(u^1_{i,j,k+1} - u^1_{i,j,k-1} + u^1_{i-1,j,k+1} - u^1_{i-1,j,k-1})$$

$$+ \frac{1}{4\Delta x_1}(u^3_{i+1,j,k} - u^3_{i-1,j,k} + u^3_{i+1,j,k-1} - u^3_{i-1,j,k-1})\big) - 2\epsilon^{13\alpha}_0 \Big]$$

$$+ c^\alpha_{16}\Big[\big(\frac{1}{4\Delta x_2}(u^1_{i,j+1,k} - u^1_{i,j-1,k} + u^1_{i-1,j+1,k} - u^1_{i-1,j-1,k})$$

$$+ \frac{1}{4\Delta x_1}(u^2_{i+1,j,k} - u^2_{i-1,j,k} + u^2_{i+1,j-1,k} - u^2_{i-1,j-1,k})\big) - 2\epsilon^{12\alpha}_0 \Big]$$

$$\bigg] h(\phi^{\alpha}_{i,j,k})$$

$$
S^{12}_{i,j,k} = \Bigg[
$$

$$
c^{\alpha}_{61}\Big(\frac{1}{4\Delta x_1}(u^1_{i+1,j,k} - u^1_{i-1,j,k} + u^1_{i+1,j+1,k} - u^1_{i-1,j+1,k}) - \epsilon^{11\alpha}_0\Big)
$$

$$
+ c^{\alpha}_{62}\Big(\frac{1}{4\Delta x_2}(u^2_{i,j+1,k} - u^2_{i,j-1,k} + u^2_{i+1,j+1,k} - u^2_{i+1,j-1,k}) - \epsilon^0_{22\alpha}\Big)
$$

$$
+ c^{\alpha}_{63}\Big(\frac{1}{4\Delta x_3}(u^3_{i,j,k} - u^3_{i,j,k-1} + u^3_{i+1,j,k} - u^3_{i+1,j,k-1}
$$

$$
+ u^3_{i,j+1,k} - u^3_{i,j+1,k-1} + u^3_{i+1,j+1,k} - u^3_{i+1,j+1,k-1}) - \epsilon^{33\alpha}_0\Big)
$$

$$
+ c^{\alpha}_{64}\Big[\Big(\frac{1}{4\Delta x_3}(u^2_{i,j,k+1} - u^2_{i,j,k-1} + u^2_{i+1,j,k+1} - u^2_{i+1,j,k-1})
$$

$$
+ \frac{1}{4\Delta x_2}(u^3_{i,j+1,k} - u^3_{i,j,k} + u^3_{i+1,j+1,k} - u^3_{i+1,j,k}
$$

$$
+ u^3_{i,j+1,k-1} - u^3_{i,j,k-1} + u^3_{i+1,j+1,k-1} - u^3_{i+1,j,k-1})\Big) - 2\epsilon^{23\alpha}_0\Big]
$$

$$
+ c^{\alpha}_{65}\Big[\Big(\frac{1}{4\Delta x_3}(u^1_{i,j,k+1} - u^1_{i,j,k-1} + u^1_{i,j+1,k+1} - u^1_{i,j+1,k-1})
$$

$$
+ \frac{1}{4\Delta x_1}(u^3_{i+1,j,k} - u^3_{i,j,k} + u^3_{i+1,j,k-1} - u^3_{i,j,k-1}
$$

$$
+ u^3_{i+1,j+1,k} - u^3_{i,j+1,k} + u^3_{i+1,j+1,k-1} - u^3_{i,j+1,k-1})\Big) - 2\epsilon^{13\alpha}_0\Big]
$$

$$
+ c^{\alpha}_{66}\Big[\Big(\frac{1}{\Delta x_2}u^1_{i,j+1,k} + \frac{1}{\Delta x_1}(u^2_{i+1,j,k} - u^2_{i,j,k})\Big) - 2\epsilon^{12\alpha}_0\Big]
$$

$$
\Bigg] h\Big(\frac{1}{4}(\phi^{\alpha}_{i,j,k} + \phi^{\alpha}_{i+1,j,k} + \phi^{\alpha}_{i,j+1,k} + \phi^{\alpha}_{i+1,j+1,k})\Big)
$$

$$
S^{12}_{i,j-1,k} = \Bigg[
$$

$$
c^{\alpha}_{61}\Big(\frac{1}{4\Delta x_1}(u^1_{i+1,j-1,k} - u^1_{i-1,j-1,k} + u^1_{i+1,j,k} - u^1_{i-1,j,k}) - \epsilon^{11\alpha}_0\Big)
$$

$$+ c_{62}^{\alpha} \Big(\frac{1}{4\Delta x_2} (u_{i,j,k}^2 - u_{i,j-2,k}^2 + u_{i+1,j,k}^2 - u_{i+1,j-2,k}^2) - \epsilon_0^{22\alpha} \Big)$$

$$+ c_{63}^{\alpha} \Big(\frac{1}{4\Delta k} (u_{i,j-1,k}^3 - u_{i,j-1,k-1}^3 + u_{i+1,j-1,k}^3 - u_{i+1,j-1,k-1}^3$$

$$+ u_{i,j,k}^3 - u_{i,j,k-1}^3 + u_{i+1,j,k}^3 - u_{i+1,j,k-1}^3) - \epsilon_0^{33\alpha} \Big)$$

$$+ c_{64}^{\alpha} \Big[$$

$$\Big(\frac{1}{4\Delta x_3} (u_{i,j-1,k+1}^2 - u_{i,j-1,k-1}^2 + u_{i+1,j-1,k+1}^2 - u_{i+1,j-1,k-1}^2)$$

$$+ \frac{1}{4\Delta x_2} (u_{i,j,k}^3 - u_{i,j-1,k}^3 + u_{i+1,j,k}^3 - u_{i+1,j-1,k}^3$$

$$+ u_{i,j,k-1}^3 - u_{i,j-1,k-1}^3 + u_{i+1,j,k-1}^3 - u_{i+1,j-1,k-1}^3) \Big)$$

$$- 2\epsilon_0^{23\alpha} \Big]$$

$$+ c_{65}^{\alpha} \Big[\Big(\frac{1}{4\Delta x_3} (u_{i,j-1,k+1}^1 - u_{i,j-1,k-1}^1 + u_{i,j,k+1}^1 - u_{i,j,k-1}^1)$$

$$+ \frac{1}{4\Delta x_1} (u_{i+1,j-1,k}^3 - u_{i,j-1,k}^3 + u_{i+1,j-1,k-1}^3 - u_{i,j-1,k-1}^3$$

$$+ u_{i+1,j,k}^3 - u_{i,j,k}^3 + u_{i+1,j,k-1}^3 - u_{i,j,k-1}^3) \Big) - 2\epsilon_0^{13\alpha} \Big]$$

$$+ c_{66}^{\alpha} \Big[\Big(- \frac{1}{\Delta x_2} u_{i,j-1,k}^1 + \frac{1}{\Delta x_1} (u_{i+1,j-1,k}^2 - u_{i,j-1,k}^2) \Big) - 2\epsilon_0^{12\alpha} \Big]$$

$$\Big] h \Big(\frac{1}{4} (\phi_{i,j-1,k}^{\alpha} + \phi_{i+1,j-1,k}^{\alpha} + \phi_{i,j,k}^{\alpha} + \phi_{i+1,j,k}^{\alpha}) \Big)$$

$$S_{i,j,k}^{13} = \Big[c_{51}^{\alpha} \Big(\frac{1}{4\Delta x_1} (u_{i+1,j,k+1}^1 - u_{i-1,j,k+1}^1 + u_{i+1,j,k}^1 - u_{i-1,j,k}^1) - \epsilon_0^{11\alpha} \Big)$$

$$+ c_{52}^{\alpha} \Big(\frac{1}{4\Delta x_2} (u_{i,j,k}^2 - u_{i,j-1,k}^2 + u_{i+1,j,k}^2 - u_{i+1,j-1,k}^2$$

$$+ u_{i,j,k+1}^2 - u_{i,j-1,k+1}^2 + u_{i+1,j,k+1}^2 - u_{i+1,j-1,k+1}^2) - \epsilon_0^{22\alpha} \Big)$$

$$+ c_{53}^{\alpha} \Big(\frac{1}{4\Delta x_3} (u_{i,j,k+1}^3 - u_{i,j,k-1}^3 + u_{i+1,j,k+1}^3 - u_{i+1,j,k-1}^3) - \epsilon_0^{33\alpha} \Big)$$

$$+ c_{54}^{\alpha} \Big[\Big(\frac{1}{4\Delta x_3} (u_{i,j,k+1}^2 - u_{i,j,k}^2 + u_{i,j-1,k+1}^2 - u_{i,j-1,k}^2$$

$$+ u_{i+1,j,k+1}^2 - u_{i+1,j,k}^2 + u_{i+1,j-1,k+1}^2 - u_{i+1,j-1,k}^2)$$

$$+ \frac{1}{4\Delta x_2}(u^3_{i,j+1,k} - u^3_{i,j-1,k} + u^3_{i+1,j+1,k} - u^3_{i+1,j-1,k})) - 2\epsilon_0^{23\alpha}\Big]$$

$$+ c^{\alpha}_{55}\Big[\Big(\frac{1}{\Delta x_3}u^1_{i,j,k+1} + \frac{1}{\Delta x_1}(u^3_{i+1,j,k} - u^3_{i,j,k})\Big) - 2\epsilon_0^{13\alpha}\Big]$$

$$+ c^{\alpha}_{56}\Big[\Big(\frac{1}{4\Delta x_2}(u^1_{i,j+1,k} - u^1_{i,j-1,k} + u^1_{i,j+1,k+1} - u^1_{i,j-1,k+1})$$

$$+ \frac{1}{4\Delta x_1}(u^2_{i+1,j,k} - u^2_{i,j,k} + u^2_{i+1,j,k+1} - u^2_{i,j,k+1}$$

$$+ u^2_{i+1,j-1,k} - u^2_{i,j-1,k} + u^2_{i+1,j-1,k+1} - u^2_{i,j-1,k+1})) - 2\epsilon_0^{12\alpha}\Big]\Big]$$

$$h(\frac{1}{4}(\phi^{\alpha}_{i,j,k} + \phi^{\alpha}_{i,j+1,k} + \phi^{\alpha}_{i,j,k+1} + \phi^{\alpha}_{i,j+1,k+1}))$$

$$S^{13}_{i,j,k-1} = \Big[$$

$$c^{\alpha}_{51}\Big(\frac{1}{4\Delta x_1}(u^1_{i+1,j,k} - u^1_{i-1,j,k} + u^1_{i+1,j,k-1} - u^1_{i-1,j,k-1}) - \epsilon_0^{11\alpha}\Big)$$

$$+ c^{\alpha}_{52}\Big(\frac{1}{4\Delta x_2}(u^2_{i,j,k-1} - u^2_{i,j-1,k-1} + u^2_{i+1,j,k-1} - u^2_{i+1,j-1,k-1}$$

$$+ u^2_{i,j,k} - u^2_{i,j-1,k} + u^2_{i+1,j,k} - u^2_{i+1,j-1,k}) - \epsilon_0^{22\alpha}\Big)$$

$$+ c^{\alpha}_{53}\Big(\frac{1}{4\Delta x_3}(u^3_{i,j,k} - u^3_{i,j,k-2} + u^3_{i+1,j,k} - u^3_{i+1,j,k-2}) - \epsilon_0^{33\alpha}\Big)$$

$$+ c^{\alpha}_{54}\Big[$$

$$\Big(\frac{1}{4\Delta x_3}(u^2_{i,j,k} - u^2_{i,j,k-1} + u^2_{i,j-1,k} - u^2_{i,j-1,k-1}$$

$$+ u^2_{i+1,j,k} - u^2_{i+1,j,k-1} + u^2_{i+1,j-1,k} - u^2_{i+1,j-1,k-1})$$

$$+ \frac{1}{4\Delta x_2}(u^3_{i,j+1,k-1} - u^3_{i,j-1,k-1} + u^3_{i+1,j+1,k-1} - u^3_{i+1,j-1,k-1}))$$

$$- 2\epsilon_0^{23}\Big]$$

$$+ c^{\alpha}_{55}\Big[\Big(-\frac{1}{\Delta x_3}u^1_{i,j,k-1} + \frac{1}{\Delta x_1}(u^3_{i+1,j,k-1} - u^3_{i,j,k-1})\Big) - 2\epsilon_0^{13}\Big]$$

$$+ c^{\alpha}_{56}\Big[\Big(\frac{1}{4\Delta x_2}(u^1_{i,j+1,k-1} - u^1_{i,j-1,k-1} + u^1_{i,j+1,k} - u^1_{i,j-1,k})$$

$$+ \frac{1}{4\Delta x_1} (u^2_{i+1,j,k-1} - u^2_{i,j,k-1} + u^2_{i+1,j,k} - u^2_{i,j,k}$$

$$+ u^2_{i+1,j-1,k-1} - u^2_{i,j-1,k-1} + u^2_{i+1,j-1,k} - u^2_{i,j-1,k})) - 2\epsilon_0^{12\alpha} \Bigg]$$

$$\Bigg] h(\frac{1}{4}(\phi^\alpha_{i,j,k-1} + \phi^\alpha_{i,j+1,k-1} + \phi^\alpha_{i,j,k} + \phi^\alpha_{i,j+1,k}))$$

The original factors of $u^1_{i,j,k}$ are gathered as a coefficient d:

$$d = \Bigg(\frac{c^\alpha_{11}}{(\Delta x_1)^2} \left(h\left(\phi^\alpha_{i+1,j,k}\right) + h\left(\phi^\alpha_{i,j,k}\right) \right)$$

$$+ \frac{c^\alpha_{66}}{(\Delta x_2)^2} \left(h\left(\frac{1}{4}(\phi^\alpha_{i,j,k} + \phi^\alpha_{i+1,j,k} + \phi^\alpha_{i,j+1,k} + \phi^\alpha_{i+1,j+1,k}) \right) \right.$$

$$\left. + h\left(\frac{1}{4}(\phi^\alpha_{i,j-1,k} + \phi^\alpha_{i+1,j-1,k} + \phi^\alpha_{i,j,k} + \phi^\alpha_{i+1,j,k}) \right) \right)$$

$$+ \frac{c^\alpha_{55}}{(\Delta x_3)^2} \left(h\left(\frac{1}{4}(\phi^\alpha_{i,j,k} + \phi^\alpha_{i,j+1,k} + \phi^\alpha_{i,j,k+1} + \phi^\alpha_{i,j+1,k+1}) \right) \right.$$

$$\left. + h\left(\frac{1}{4}(\phi^\alpha_{i,j,k-1} + \phi^\alpha_{i,j+1,k-1} + \phi^\alpha_{i,j,k} + \phi^\alpha_{i,j+1,k}) \right) \right) \Bigg).$$

Symmetry considerations and explicit writing of the equations for the components $\mathbf{u}_2$ and $\mathbf{u}_3$ in cell (i,j,k) show that their update schemes can by gained by simply 'renaming' the indexes of the components that enter the update for $\mathbf{u}_1$. So, the exact same scheme as for the update of component $\mathbf{u}_1$ can be used for the update of $\mathbf{u}_2$ and $\mathbf{u}_3$ just by renaming index positions. The exact correspondence of the coefficients is shown explicitly in Tab. 7.1. As can be seen, the schemes differ only by a transposition of two indexes, while the third one stays fixed. In the appendix, a more formal version of this interrelation, based on the tensor notation, is given (see Appendix B.3). Writing down the equations for all three components of $\mathbf{u}$ in all non-boundary cells results in a system of linear equations. This system can be solved iterative using a Gauß-Seidel algorithm that can be combined with a *successive over-relaxation* (SOR) method (see [107]) and [101]). To parallelize the scheme, a red-black variant is applied, where the grid is thought to be colored alternating checkerboard-like (in 3D) in red and black. To update the field, first the red points are updated, and then the black ones. This strategy permits the parallelization of the

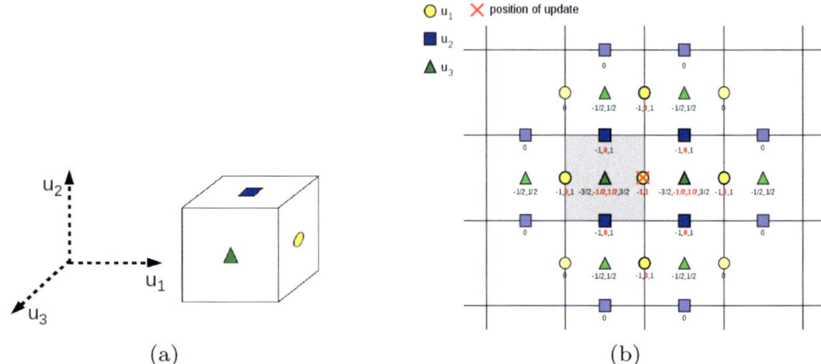

(a)　　　　　　　　　　　　　　　　　(b)

Figure 7.1: **a)** Staggered grid discretization scheme for $\mathbf{u}_1$, $\mathbf{u}_2$ and $\mathbf{u}_3$ shown for a single cell. **b)** Sketch of the needed displacement components for the update of component $\mathbf{u}_1$ in a specific cell (marked with a red cross). The illustration shows a plane through $\mathbf{u}_1$ and $\mathbf{u}_2$, such that the grid of component $\mathbf{u}_3$ lies outside this plane. The numbers and fractions give the distance of the needed values relative to the plane.

scheme giving the same result as the single-core implementation, if the elastic stiffness matrix C has only non-zero entries in the components C_{ii} and C_{jk} for $i = 1, \ldots, 6$ and $j, k = 1, 2, 3$, because in this case red-colored grid points only depend o n bla ck ones and vise versa. If other entries of the stiffness tensor are non-zero (e.g. because arbitrary orientations are allowed or low material symmetries are applied), the red-black scheme does not compute the same values everywhere in the domain in single-core and multi-core simulations.[4]

7.3.2 Boundary conditions

Boundary conditions for the elastic problems are more complex than the ones in the case of the phase-fields, because mostly boundary conditions

---

[4] As a domain decomposition scheme is applied to parallelize the computations, the differences are induced at the 'boundaries' of the part of Ω that is treated by a single process, as boundary values have to be exchanged across processes.

for RVEs are needed where the periodicity for the strain has to be maintained. The most relevant boundary conditions for micromechanic simulations are given here. For this discussion let $\partial\Omega$ be the boundary of the calculation domain Ω, $\omega \in \partial\Omega$ a point on the boundary and n the normal on the boundary pointing outwards.

Dirichlet boundary condition for the displacement components The displacement field values are fixed on the boundary: $\mathbf{u}(\omega) = c(\omega) \in \mathbb{R}^3$. The implementation has to respect the staggered grid structure and is described in [100]. The special case of

$$\mathbf{u}_{|\partial\Omega} \equiv \mathbf{0}$$

is called the *clamped boundary condition* in the literature (see e.g. [94]). In combination with eigenstrains, this boundary condition might be applied after the initial structure is relaxed under e.g. a stress-free boundary condition.

Constant traction boundary condition The traction forces at the boundary in the direction of n are fixed at the boundaries, allowing to represent stress controlled experiments

$$\boldsymbol{\sigma}_{|\partial\Omega}\, n(\omega) = \left(\sum_{\alpha=1}^{N} h(\phi_\alpha)\mathcal{C}^\alpha(\boldsymbol{\epsilon} - \boldsymbol{\epsilon}_0^\alpha) \right) n(\omega) = c(\omega) \in \mathbb{R}^3.$$

Due to the staggered grid discretization, the normal components at the boundaries lie half a cell width displaced inside the domain, what increases the discretization error near the boundary. A first implementation of this boundary condition for cubic and an anisotropic materials was implemented in [100], and this implementation was generalized to consider materials providing up to simple tetragonal or hexagonal symmetry (see Appendix B.2). The special case of

$$\boldsymbol{\sigma}_{|\partial\Omega} \equiv \mathbf{0} \tag{7.7}$$

is called the *free boundary condition*.

Periodic strains To gain periodic boundary conditions that represent RVEs, the following condition (cp. the book of Nemat-Nasser [108] or the PhD thesis of A. Fröhlich [109]) is implemented:

$$\mathbf{u}(x^+) - \mathbf{u}(x^-) = \bar{\epsilon}(x^+ - x^-), \tag{7.8}$$

where x^+ and x^- are points at opposite boundaries, and $\bar{\epsilon}$ is a homogeneous strain imposed to the system. The idea of the decomposition of strain as proposed by Khachaturyan [110, 94] is assumed:

$$\epsilon = \bar{\epsilon} + \delta\epsilon, \tag{7.9}$$

where $\bar{\epsilon}$ is related to the homogeneous strain of the system (i.e. the change of shape and volume), and the heterogeneous strain satisfies $\int_\Omega \delta\epsilon = 0$. The *totally clamped boundary condition* is reflected by $\bar{\epsilon} = \mathbf{0}$ (see [94]). To account for an external applied stress $\boldsymbol{\sigma}^{\mathrm{appl}}$, an approach according to Wu et al. is used (see [94]). The *potential energy*, i.e. the difference between the elastic energy and the work the systems performs against the applied stress, is defined as

$$E_p = E_{\mathrm{elast}} - |\Omega|\boldsymbol{\sigma}^{\mathrm{appl}}\bar{\epsilon}.$$

An expression for $\bar{\epsilon}$ can now be gained by extremizing E_p with respect to $\bar{\epsilon}$. Taking Eq. (7.9) into account, the elastic energy reads

$$f_{\mathrm{elast}} = \frac{1}{2}\sum_{\alpha=1}^{N} h(\phi_\alpha)(\bar{\epsilon} + \delta\epsilon - \epsilon_0^\alpha) \cdot C^\alpha(\bar{\epsilon} + \delta\epsilon - \epsilon_0^\alpha).$$

Then

$$\frac{\partial E_{\mathrm{elast}}}{\partial\bar{\epsilon}} = \frac{\partial}{\partial\bar{\epsilon}}\int_\Omega \frac{1}{2}\sum_{\alpha=1}^{N} h(\phi_\alpha)(\bar{\epsilon} + \delta\epsilon - \epsilon_0^\alpha) \cdot C^\alpha(\bar{\epsilon} + \delta\epsilon - \epsilon_0^\alpha)\, d\Omega$$

$$= \int_\Omega \sum_{\alpha=1}^{N} h(\phi_\alpha)C^\alpha\bar{\epsilon}\, d\Omega + \int_\Omega \sum_{\alpha=1}^{N} h(\phi_\alpha)C^\alpha\delta\epsilon\, d\Omega -$$

$$\int_\Omega \sum_{\alpha=1}^{N} h(\phi_\alpha)C^\alpha\epsilon_0^\alpha\, d\Omega.$$

The last equality follows from the symmetry ofthe tensors ϵ and the ϵ_0^α. Assuming for all phases the same elastic stiffness, i.e. $C^\alpha = C_{\mathrm{hom}}$ for all $\alpha \leq N$, and $\sum_{\alpha=1}^{N} h(\phi_\alpha) \equiv 1$, the above expression can be simplified to

$$\frac{\partial E_{\mathrm{elast}}}{\partial \bar{\epsilon}} = C_{\mathrm{hom}}\bar{\epsilon} \int_\Omega \mathrm{d}\Omega - C_{\mathrm{hom}} \int_\Omega h(\phi_\alpha)\epsilon_0^\alpha \, \mathrm{d}\Omega$$

$$= |\Omega|C_{\mathrm{hom}}\bar{\epsilon} - C_{\mathrm{hom}} \int_\Omega h(\phi_\alpha)\epsilon_0^\alpha \, \mathrm{d}\Omega,$$

as $\bar{\epsilon}$ is constant and $\int_\Omega \delta\epsilon = 0$. Because

$$\frac{\partial}{\partial \bar{\epsilon}} |\Omega|\sigma^{\mathrm{appl}}\bar{\epsilon} = |\Omega|\sigma^{\mathrm{appl}},$$

the relation

$$\frac{\partial E_p}{\partial \bar{\epsilon}} = |\Omega|C_{\mathrm{hom}}\bar{\epsilon} - C_{\mathrm{hom}} \int_\Omega \sum_{\alpha=1}^{N} h(\phi_\alpha)\epsilon_0^\alpha \, \mathrm{d}\Omega - |\Omega|\sigma^{\mathrm{appl}}$$

holds. From this and $\frac{\partial E_p}{\partial \bar{\epsilon}} = 0$, $\bar{\epsilon}$ can be calculated as

$$\bar{\epsilon} = \frac{1}{|\Omega|} \int_\Omega \sum_{\alpha=1}^{N} h(\phi_\alpha)\epsilon_0^\alpha \, \mathrm{d}\Omega + C_{\mathrm{hom}}^{-1}\sigma^{\mathrm{appl}}. \tag{7.10}$$

If heterogeneous or arbitrarily oriented elastic properties are allowed, the equation for $\bar{\epsilon}$ becomes much more complicated. In this case the interpolation of the stiffness matrices C^α have to be calculated, and the inverse of this interpolated matrix, what is computational demanding.

7.3.3 Vefication: The Eshelby inclusion problem

To verify the correctness of the implementation of the SOR method that is used to calculate the mechanical equilibrium (see the end of Sec. 7.3.1), a scenario representing a version of *Eshelby's inclusion problem* was set-up. Eshelby derived an analytic expression for the stress and strain distribution of a homogeneous inclusion in an infinite matrix (see [43]). The parameters of the numerical experiment are taken from a work of Apel and

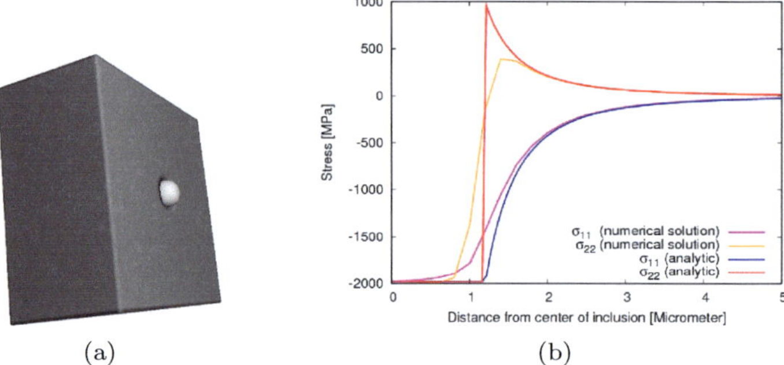

(a)	(b)

Figure 7.2: Simulation of an inclusion of low-alloyed steel in a matrix: **(a)** Sketch of the inclusion in the matrix and **(b)** comparison of the stress profile measured from the center of the inclusion compared to the analytical solution. The main deviation appears in the region of the diffusive interface. Similar numerical experiments were carried out by Apel and Steinbach [111].

Steinbach [111], and reflect a low alloyed steel (represented by a Youngs modulus of 280 GPa and a Poission ratio of 0.3). The inclusion of size $1.2\mu m$ has a uniform eigenstrain of 1%, the surrounding matrix is undistorted. The total size of the simulation box is $15\mu m$. The boundary conditions for the phase-fields are of the special Neumann-type, for the elastic displacement field the free boundary condition Eq. (7.7) is applied. Fig. 7.2 shows the initial problem and the resulting strain components compared to the analytical expression. The results compare quite well with the analytic solution and show differences in the interfacial region. The width of the diffuse interface is resolved by five grid points what corresponds to a physical width of about $1\mu m$. The phase fields representing the inclusion and the matrix in this simulations were held fixed.

7.4 Discretization of the micromagnetic contributions

The micromagnetic equations are discretized on the collocated grid. The interpretation of the vector field $\mathbf{m}$ of spontaneous magnetization differs from the interpretation of the physical quantities so far. The value of $\mathbf{m}$ at a specific grid point is assumed to be a volume averaged quantity over the cell with volume $\Delta x \Delta y \Delta z$, and $\mathbf{m}$ is the representative in the center of the cell. Therefore, the specimen represented by the domain Ω is thought to be partitioned into $N \in \mathbb{N}$ equisized parallel-epipedic cells $\Omega_i \subseteq \Omega$ $(i = 1, \ldots, N)$ with

$$\Omega = \dot{\bigcup} \Omega_i$$

and for all $i, j, \leq N$ with $i \neq j$

$$\Omega_i \cap \Omega_j = \emptyset \quad \text{and} \quad |\Omega_i| = |\Omega_j|.$$

Due to the interpretation of $\mathbf{m}$ as a volume averaged quantity, in each cell Ω_i the magnetization vector $\mathbf{m}$ is assumed to be constant. Hence, there are functions

$$C_i : \mathbb{R}_{\geq 0} \to \mathbb{S}^2$$

such that

$$\mathbf{m}_{|\Omega_i}(t) \equiv C_i(t).$$

All arising spatial derivatives are approximated using finite differences techniques. This is done in agreement with the work published by Miltat and Donahue [52].

The micromagnetic boundary conditions are implemented analogously to the boundaries of the phase fields: For finite extended specimens, the condition $\frac{\partial \mathbf{m}}{\partial n} = 0$ is appropriate, and for periodically extended magnetization states as used for RVEs, the periodic boundary condition is used (see [52]). The computational demanding calculation of the demagnetization field (cp. Sec. 4.2) needs special numerical treatment to make simulations feasible. Furthermore, care must be taken to numerically compute the dynamics of the spontaneous magnetization using the Landau-Lifshitz-Gilbert equation Eq. (4.8). A more detailed discussion on the computation of the micromagnetic demagnetization field and the evolution solution procedures is given in the next chapter.

Table 7.1: The table explicitly shows the transposition rules when the update of the displacement field components $\mathbf{u}_2$ and $\mathbf{u}_3$ are computed, using the exact same scheme as discussed for $\mathbf{u}_1$ but by renaming the indexes. The relation of the elastic stiffness coefficients c_{ij} and the components of the stress and strain tensors in computing the update of the components $\mathbf{u}_m$ ($m = 1, 2, 3$) is shown for the computation of a mechanical equilibrium. The update scheme for $\mathbf{u}_1$ is fixed as a reference scheme. For the sake of better readability, the phase index α is not shown.

Update of $\mathbf{u}_1$	Update of $\mathbf{u}_2$	Update of $\mathbf{u}_3$
x_1	x_2	x_3
x_2	x_1	x_2
x_3	x_3	x_1
σ_{11}	σ_{22}	σ_{33}
σ_{12}	σ_{21}	σ_{32}
σ_{13}	σ_{23}	σ_{31}
ϵ_{11}	ϵ_{22}	ϵ_{33}
ϵ_{22}	ϵ_{11}	ϵ_{22}
ϵ_{33}	ϵ_{33}	ϵ_{11}
$\epsilon_{23} = \epsilon_{32}$	$\epsilon_{13} = \epsilon_{31}$	$\epsilon_{21} = \epsilon_{12}$
$\epsilon_{13} = \epsilon_{31}$	$\epsilon_{23} = \epsilon_{32}$	$\epsilon_{31} = \epsilon_{13}$
$\epsilon_{12} = \epsilon_{21}$	$\epsilon_{21} = \epsilon_{12}$	$\epsilon_{32} = \epsilon_{23}$
c_{11}	c_{22}	c_{33}
c_{12}	c_{21}	c_{32}
c_{13}	c_{23}	c_{31}
c_{14}	c_{25}	c_{36}
c_{15}	c_{24}	c_{35}
c_{16}	c_{26}	c_{34}
c_{61}	c_{62}	c_{43}
c_{62}	c_{61}	c_{42}
c_{63}	c_{63}	c_{41}
c_{64}	c_{65}	c_{46}
c_{65}	c_{64}	c_{45}
c_{66}	c_{66}	c_{44}
c_{51}	c_{42}	c_{53}
c_{52}	c_{41}	c_{52}
c_{53}	c_{43}	c_{51}
c_{54}	c_{45}	c_{56}
c_{55}	c_{44}	c_{55}
c_{56}	c_{46}	c_{54}

8 Micromagnetic evolution

The Landau-Lifshitz-Gilbert Eq. (4.8), described in Chap. 4, is the widely accepted equation for the time-evolution of the magnetic moments in ferromagnetic materials. There are some major issues that must be dealt with when a solution method is numerically implemented. The magnetic moments locally precede around the axis of a so called effective magnetic field, and in the equilibrium state the directions of the effective magnetic field and the local moments locally coincide. The demagnetization field, arising from the interdependent interaction of the magnetic moments, is a non-negligible addend of this effective field. The long-range character of the demagnetization field makes it hard to compute. This chapter deals with two major topics: The first section discusses problems with the numerical integration of Eq. (4.8) and summarizes an unconditional stable explicit one-step Euler integration scheme that will be used to compute the updates as proposed by Lewis and Nigam [112, 12]. The second section discusses the difficulties in conjunction with the calculation of the demagnetization field. Two solution schemes will be presented to calculate the demagnetization field efficiently for different boundary conditions, both will rely on FFT techniques: One solution method assumes a finitely extended specimen, the other an in all three spatial dimensions periodic RVE, cut out of a surrounding specimen.

8.1 Time-integration of the Landau-Lifshitz-Gilbert equation

This section discusses the problems that arise when the Landau-Lifshitz-Gilbert equation Eq.(4.8) is solved. Some drawbacks of conventional integration schemes are pointed out, and an unconditional norm conservative scheme that was published by Lewis and Nigam [12, 112] is presented.

8.1.1 Classical integration schemes and the Landau-Lifshitz-Gilbert equation

To simplify the following discussion, the consideration is limited to a single magnetic moment at a fixed point ω in the domain Ω, so $\mathbf{m} : \mathbb{R}_{\geq 0} \to \mathbb{S}^2$ becomes a mere function of time t. Assume a discrete time update scheme of the form

$$\mathbf{m}(t_n + \Delta t) = \mathbf{m}(t_n) + F(t_n, \Delta t, \mathbf{m}(t_n)) \tag{8.1}$$

to compute the updates, where Δt is the discrete time-step width and F the function that describes the update rule. Independent of how the update function F looks exactly, this scheme describes a translation of $\mathbf{m}^n = \mathbf{m}(t_n)$ when moving from time t_n to time $t_{n+1} = t_n + \Delta t$. Therefore $|\mathbf{m}^{n+1}| = 1$ is not assured, i.e. $\mathbf{m}$ is not unconditionally enforced to stay on the unit sphere. Special techniques are needed to ensure that constraint with such classical schemes (see [30] for a more general discussion of this fact).

An obvious way to enforce $|\mathbf{m}(t_n)| = 1$ during numerical integration is *explicit renormalization* of the field variable after the update step. If $|\mathbf{m}(t_n)| \neq 1$, then $|\mathbf{m}(t_n)|$ is renormalized by projecting the update onto a valid solution:

$$\frac{\mathbf{m}(t_n)}{|\mathbf{m}(t_n)|}.$$

The example and reasoning presented now follow an article by Lewis and Nigam [112]. There are two main issues with the act of the explicit renormalization: First, it is 'aphysical' in the sense that it means adding (or subtracting) energy to (or from) a system. And secondly, renormalization may change the potential ψ that describes the demagnetization interactions $\mathbf{H}_{\mathrm{demag}} = -\nabla \psi$ (cp. Sec. 4.1 and Eq. (4.2)) in a non-linear way by affecting the divergence: Assume a planar magnetization state in the standard basis on $\mathbb{R}^3$ given by

$$\mathbf{m}(x, y, z) = a\mathbf{x}_1 + b(y)\mathbf{x}_2,$$

where $a \in \mathbb{R}$ is a constant, $b : \mathbb{R} \to \mathbb{R}$ is a scalar function. Furthermore, assume a discrete scheme that computes $\mathbf{m}^{n+1}(x, y, z)$ such that $\mathbf{m}^{n+1} \notin \mathbb{S}^2$, but

$$\nabla \cdot \mathbf{m}^{n+1} = 0.$$

The renormalized field is then, in general, not divergence-free:

$$\nabla \cdot \left(\frac{\mathbf{m}^{n+1}(x, y, z)}{|\mathbf{m}^{n+1}(x, y, z|} \right) = -\frac{ab(x)b'(x)}{(a^2 + b(x)^2)^{\frac{3}{2}}} \underset{\text{in general}}{\neq} 0.$$

Here, b' denotes the derivation of the scalar function b with respect to its only argument. As $\mathbf{m}$ enters into the computation of the demagnetization field and energy after the update, the obtained solution is altered by the act of renormalization.

A second problem with classical schemes has been pointed out by Wang, E and García [113]. They performed a stability analysis of the integration of the Eq. (4.8) using an explicit Euler scheme as in Eq. (8.1). Assuming the Eq. (4.8) to be of the simple form

$$\dot{\mathbf{m}} = -a \times \mathbf{m},$$

with a constant $a = (a_1, a_2, a_3)^T \in \mathbb{R}^3$, the explicit updates take the form

$$\mathbf{m}^{n+1} = \mathbf{m}^n - \Delta t(a \times \mathbf{m}^n).$$

In matrix-vector notation this reads

$$\mathbf{m}^{n+1} = A(\Delta t)\mathbf{m}^n,$$

with a matrix

$$A(\Delta t) = \mathbf{I} - \Delta t \operatorname{skew}(a) = \begin{pmatrix} 1 & \Delta t a_3 & -\Delta t a_2 \\ -\Delta t a_3 & 1 & \Delta t a_1 \\ \Delta t a_2 & -\Delta t a_1 & 1 \end{pmatrix}.$$

For analyzing the stability of this Euler scheme, one needs the eigenvalues of the problem matrix $A(\Delta t)$. These are the roots of the characteristic polynomial of A:

$$\det(A - \lambda I) = (1 - \lambda)^3 + (1 - \lambda)(a_1^2 + a_2^2 + a_3^2)(\Delta t)^2.$$

The roots are

$$\lambda_0 = 1, \lambda_1 = 1 + i|a|\Delta t, \lambda_2 = 1 - i|a|\Delta t,$$

where i denotes the complex imaginary unit. The spectral radius ρ of A is the supremum of the absolute values of the eigenvalues

$$\rho(A) = \sqrt{1 + |a|^2(\Delta t)^2}.$$

In the relevant cases where $a \neq (0,0,0)^T$ and $\Delta t > 0$, the eigenvalues λ_1 and λ_2 are non-real complex numbers, and $\rho(A) > 1$. But for an time-integration scheme to be be stable the spectral radius of the problem matrix has to be less than one, which proves the above scheme to be in general unstable.

8.1.2 An unconditional stable explicit one-step geometric integration scheme

A natural idea to overcome the issues of the last section is to use rotations $R \in \mathrm{SO}(3)$ to compute updates of the form

$$\mathbf{m}^{n+1} = R^n(\Delta t, \mathbf{m}^n).$$

As rotations are Euclidean motions (cp. Sec. 2.3), they do not alter the length of a vector and therefore unconditionally fulfill the constraint $\mathbf{m}^{n+1} \in \mathbb{S}^2$ at all times without the need of problematical projection procedures.

The results from Lie-group theory (see Sec. 2.5) will be used to explain the unconditionally norm-conservative one-step time-integration scheme published by Lewis and Nigam [12, 112]. Many ideas presented by Iserles et al. [30] enter here. The basic idea is to rewrite Eq. (4.8) in terms of a transitive (and non-free) Lie-group action on a manifold, and then apply Thm. 2.13 to gain an integration scheme with the desired properties. The goal is to find a path on the unit sphere that describes the motion of a single magnetization vector over time towards equilibrium. Again, a magnetization vector $\mathbf{m}$ is fixed at $x \in \Omega$, and its evolution over time is followed. First, a continuous function $\mathbf{m} : \mathbb{R}_{\geq 0} \to \mathbb{S}^2$ is assumed, before a discretized version is considered.

Starting point are some observations from Sec. 2.5: Because $SO(3)$ acts transitively on $\mathbb{S}^2$, a time independent constant start magnetization $\mathbf{m}_0 \in \mathbb{S}^2$ can be fixed and a time dependent smooth curve

$$Q : \mathbb{R}_{\geq 0} \to SO(3),$$

exists, such that

$$\mathbf{m}(t) = Q(t)\mathbf{m}_0.$$

For better readability, the arguments of the fields will be suppressed. Keep in mind that $\mathbf{m}(t)$ depends on time, and the effective field $\mathbf{H}_{\text{eff}}(\mathbf{m})$ depends on the magnetization (cp. Chap. 4.3). Differentiation of $\mathbf{m}$ with respect to the time parameter t gives

$$\dot{\mathbf{m}} = \dot{Q}\mathbf{m}_0, \tag{8.2}$$

as $\mathbf{m}_0$ is a constant and does not depend on time. Eq. (4.8) can be rewritten as

$$\dot{\mathbf{m}} = -\frac{\gamma}{(1 + \alpha_G^2)} \left(\mathbf{m} \times \mathbf{H}_{\text{eff}} + \alpha_G \mathbf{m} \times (\mathbf{m} \times \mathbf{H}_{\text{eff}}) \right) \tag{8.3}$$

$$= \frac{\gamma}{(1 + \alpha_G^2)} \left(\mathbf{H}_{\text{eff}} \times \mathbf{m} + (\alpha_G \mathbf{m} \times \mathbf{H}_{\text{eff}}) \times \mathbf{m} \right)$$

$$= \omega \times \mathbf{m}$$

$$= \text{skew}(\omega)\mathbf{m}$$

$$= \text{skew}(\omega)Q\mathbf{m}_0,$$

with the abbreviation

$$\omega = \frac{\gamma}{(1 + \alpha_G^2)} \left(\mathbf{H}_{\text{eff}} + (\alpha_G \mathbf{m} \times \mathbf{H}_{\text{eff}}) \right).$$

The Eqs. (8.2) and (8.3) show

$$\dot{Q}\mathbf{m}_0 = \text{skew}(\omega)Q\mathbf{m}_0,$$

and hence

$$\dot{Q} = \text{skew}(\omega)Q.$$

As $Q \in \mathrm{SO}(3)$ and $\mathrm{skew}(\omega) \in \mathfrak{so}(3)$, the curve Q has a right trivialization. From theorem Th. 2.13 follows

$$Q(t) = \exp(t \cdot \mathrm{skew}(\omega)).$$

Discretizing the Eq. (8.3) and writing it for time-step $n + 1$ gives

$$\mathbf{m}^{n+1} = \mathrm{skew}(\omega_n) Q_n \mathbf{m}_0$$

with

$$\omega_n = \frac{\gamma}{(1 + \alpha_G^2)} \left(\mathbf{H}_{\mathrm{eff}}^n + (\alpha_G \mathbf{m}^n \times \mathbf{H}_{\mathrm{eff}}^n) \right).$$

The solution is now given by

$$\mathbf{m}^{n+1} = \exp(\Delta t \cdot \mathrm{skew}(\omega_n)) \mathbf{m}^n.$$

A choice for the exponential map exp is essential to maintain numerical accuracy and stability. Here, the Cayley transform and the true matrix exponential (see Def. 2.29 and the end of Sec. 2.5) are good choices. Having $\omega_n \in \mathbb{R}^3$ calculated, then from the application of the Cayley transform the explicit update formula can be derived as

$$\mathbf{m}^{n+1} = \mathrm{cay}(\Delta t \, \mathrm{skew}(\omega_n)) \mathbf{m}^n \tag{8.4}$$

$$= \mathbf{m}^n + \frac{1}{1 + |\frac{1}{2} \Delta t \omega_n|^2} \left(\Delta t \omega_n \times \mathbf{m}^n + \frac{1}{2} \Delta t \omega_n \times (\Delta t \omega_n \times \mathbf{m}^n) \right),$$

as an explicit update scheme for $\mathbf{m}$. The solution scheme using the true exponential becomes (by applying angle-doubling formulae as in [30, Appendix B])

$$\mathbf{m}^{n+1} = \exp(\Delta t \omega_n) \mathbf{m}^n$$

$$= \mathbf{m}^n + \frac{\sin(|\Delta t \omega_n|)}{|\Delta t \omega_n|} \Delta t \omega_n \times \mathbf{m}^n + \frac{1 - \cos(|\Delta t \omega_n|)}{|\Delta t \omega_n|^2} (\Delta t \omega_n \times (\Delta t \omega_n \times \mathbf{m}^n))$$

The matrix exponential needs trigonometric functions to be evaluated, so updates for the spontaneous magnetization $\mathbf{m}$ are computed using the Cayley transform Eq. (8.4).

8.2 Calculation of the demagnetization field

The magnetostatic demagnetization energy is a crucial contribution to the energetics of micromagnetic systems. The calculation of the demagnetization field is a computationally demanding task. When the demagnetization field $\mathbf{H}_{\mathrm{demag}}$ is calculated, two significantly different assumptions concerning the extension of the computation domain Ω are made in this work: Either the specimen under consideration has a finite extension in all three spatial dimensions, or the domain Ω represents a 3D periodic RVE that is embedded in a surrounding specimen. Proposals for boundary conditions that reflect periodicity in 1D or 2D exist in the literature (see Lebecki et al. [114] or Wang et al. [115]), but they are not needed in this work.

8.2.1 Finite extended specimens

Following strictly the article of Miltat and Donahue [52], the presented calculation of the demagnetization energy and the demagnetization field for a finitely extended specimens is based on an energy-based approach. With the assumptions to the discretization for the spontaneous magnetization made in Sec. 7.4, $\mathbf{m}$ is a piecewise constant function on Ω and consequently, $\nabla \cdot \mathbf{m}_{|\Omega_i} = 0$. So, the magnetization inside the volume element Ω_i is divergence-free. The magnetostatic self energy reads (see Chap. 4)

$$E_{\mathrm{demag}} = -\frac{1}{2}\mu_0 M_s \int_{\Omega} (\mathbf{m} \cdot \mathbf{H}_{\mathrm{demag}})\, \mathrm{d}\,\Omega, \qquad (8.5)$$

and the demagnetization field $\mathbf{H}_{\mathrm{demag}}$ has the solution (cp. Eq. 4.2)

$$\mathbf{H}_{\mathrm{demag}}(\mathbf{r}) = -\frac{1}{4\pi} M_s \int_{\Omega} \nabla \cdot \mathbf{m}(\mathbf{r}') \frac{\mathbf{r} - \mathbf{r}'}{|\mathbf{r} - \mathbf{r}'|^3} \mathrm{d}^3 \mathbf{r}'$$
$$+ \frac{1}{4\pi} M_s \int_{\partial\Omega} \hat{\mathbf{n}}(\mathbf{r}') \cdot \mathbf{m}(\mathbf{r}') \frac{\mathbf{r} - \mathbf{r}'}{|\mathbf{r} - \mathbf{r}'|^3} \mathrm{d}^2 \mathbf{r}',$$

where $\partial\Omega$ denotes the surface boundary of Ω and $\hat{\mathbf{n}}$ the field of surface normals (pointing outwards). To simplify the following arguments, the function

$$g : \Omega \to \mathbb{R}^3, \mathbf{x} \mapsto \frac{1}{|\mathbf{x}|^3}\mathbf{x}$$

is introduced. g has the three component functions

$$g_i : \Omega \to \mathbb{R}, \mathbf{x} \mapsto \frac{1}{|\mathbf{x}|^3}\mathbf{x}_i, \qquad i = 1, 2, 3.$$

So, the demagnetization field can be written as

$$\mathbf{H}_{\text{demag}}(\mathbf{r}) = -\frac{1}{4\pi}M_s \int_{\Omega} \nabla \cdot \mathbf{m}(\mathbf{r}')g(\mathbf{r} - \mathbf{r}')\mathrm{d}^3\mathbf{r}'$$
$$+ \frac{1}{4\pi}M_s \int_{\partial\Omega} \hat{\mathbf{n}}(\mathbf{r}') \cdot \mathbf{m}(\mathbf{r}')g(\mathbf{r} - \mathbf{r}')\mathrm{d}^2\mathbf{r}'.$$

$\mathbf{H}_{\text{demag}}$ has the three components ($i = 1, 2, 3$)

$$(\mathbf{H}_{\text{demag}})_i(\mathbf{r}) = -\frac{1}{4\pi}M_s \int_{\Omega} \nabla \cdot \mathbf{m}(\mathbf{r}')g_i(\mathbf{r} - \mathbf{r}')\mathrm{d}^3\mathbf{r}'$$
$$+ \frac{1}{4\pi}M_s \int_{\partial\Omega} \hat{\mathbf{n}}(\mathbf{r}') \cdot \mathbf{m}(\mathbf{r}')g_i(\mathbf{r} - \mathbf{r}')\mathrm{d}^2\mathbf{r}'.$$

Applying the 'integration by parts' method on the second addend gives

$$\frac{1}{4\pi}M_s \int_{\partial\Omega} g_i(\mathbf{r} - \mathbf{r}')\hat{\mathbf{n}}(\mathbf{r}') \cdot \mathbf{m}(\mathbf{r}')\mathrm{d}^2\mathbf{r}' = \frac{1}{4\pi}M_s \int_{\Omega} g_i(\mathbf{r} - \mathbf{r}')\nabla \cdot \mathbf{m}(\mathbf{r}')\mathrm{d}^3\mathbf{r}' +$$
$$\frac{1}{4\pi}M_s \int_{\Omega} \nabla g_i(\mathbf{r} - \mathbf{r}') \cdot \mathbf{m}(\mathbf{r}')\mathrm{d}^3\mathbf{r}'.$$

Thus, the i-th component of the demagnetization field becomes

$$(\mathbf{H}_{\text{demag}})_i(\mathbf{r}) = \frac{1}{4\pi}M_s \int_{\Omega} \nabla g_i(\mathbf{r} - \mathbf{r}') \cdot \mathbf{m}(\mathbf{r}')\mathrm{d}^3\mathbf{r}'.$$

Until here, only the commutativity of $(\mathbb{R}, \cdot)$ and the standard scalar product on $\mathbb{R}^3$ were used. In a Cartesian coordinate system, the gradient ∇ operates component-wise on a vector field, hence

$$\mathbf{H}_{\text{demag}}(\mathbf{r}) = \frac{1}{4\pi}M_s \int_{\Omega} \nabla g(\mathbf{r} - \mathbf{r}')\mathbf{m}(\mathbf{r}')\mathrm{d}^3\mathbf{r}'$$

$$= \frac{1}{4\pi} M_s \sum_{i=1}^{N} \int_{\Omega_i} \nabla g(\mathbf{r} - \mathbf{r}')\mathbf{m}(\mathbf{r}')\mathrm{d}^3\mathbf{r}'.$$

Denoting by $\mathbf{r}_i$ the vector to the midpoint of the cell Ω_i, the assumption of constant magnetization in each cell reads

$$\mathbf{m}(\mathbf{r}) = \mathbf{m}(\mathbf{r}_i) \text{ for all } \mathbf{r} \in \Omega_i. \tag{8.6}$$

This discretization is second order accurate (cp. [52]). Integrating by parts again results in

$$\int_{\Omega_i} \nabla g(\mathbf{r} - \mathbf{r}')\mathbf{m}(\mathbf{r}'_i)\mathrm{d}^3\mathbf{r}' = \int_{\partial\Omega_i} \hat{\mathbf{n}}(\mathbf{r}') \cdot \mathbf{m}(\mathbf{r}')g(\mathbf{r} - \mathbf{r}')\mathrm{d}^2\mathbf{r}'$$
$$- \int_{\Omega_i} \nabla \cdot \mathbf{m}(\mathbf{r}')g(\mathbf{r} - \mathbf{r}')\mathrm{d}^3\mathbf{r}',$$

and $\int_{\Omega_i} \nabla \cdot \mathbf{m}(\mathbf{r}')g(\mathbf{r} - \mathbf{r}')\mathrm{d}^3\mathbf{r}' = 0$, because of Eq. (8.6). Thus,

$$\int_{\Omega_i} \nabla g(\mathbf{r} - \mathbf{r}')\mathbf{m}(\mathbf{r}'_i)\mathrm{d}^3\mathbf{r}' = \int_{\partial\Omega_i} \hat{\mathbf{n}}(\mathbf{r}') \cdot \mathbf{m}(\mathbf{r}')g(\mathbf{r} - \mathbf{r}')\mathrm{d}^2\mathbf{r}',$$

and a second order accurate approximation of the demagnetization field is given by

$$\mathbf{H}_{\text{demag}}(\mathbf{r}) = \frac{1}{4\pi} M_s \sum_{i=1}^{N} \int_{\partial\Omega_i} \mathbf{m}(\mathbf{r}'_i) \cdot \hat{\mathbf{n}}g(\mathbf{r} - \mathbf{r}')\mathrm{d}^2\mathbf{r}' + \mathcal{O}(\Delta^2). \tag{8.7}$$

Eq. (8.7) is inserted into the energy expression Eq. (8.5)

$$- \frac{1}{2}\mu_0 M_s \int_{\Omega} (\mathbf{m} \cdot \mathbf{H}_{\text{demag}})\,\mathrm{d}\,\Omega$$

$$\approx - \frac{1}{8\pi}\mu_0 M_s^2 \int_{\Omega} \mathbf{m}(\mathbf{r}) \cdot \left[\sum_{j=1}^{N} \int_{\partial\Omega_j} \mathbf{m}(\mathbf{r}'_j) \cdot \hat{\mathbf{n}}(\mathbf{r}')g(\mathbf{r} - \mathbf{r}')\mathrm{d}^2\mathbf{r}' \right]\mathrm{d}^3\mathbf{r}$$

$$\approx - \frac{1}{8\pi}\mu_0 M_s^2 \sum_{i=1}^{N} \int_{\Omega_i} \mathbf{m}(\mathbf{r}_i) \cdot \sum_{j=1}^{N} \int_{\partial\Omega_j} \mathbf{m}(\mathbf{r}'_j) \cdot \hat{\mathbf{n}}(\mathbf{r}')g(\mathbf{r} - \mathbf{r}')\mathrm{d}^2\mathbf{r}'\mathrm{d}^3\mathbf{r}$$

$$= - \frac{1}{8\pi}\mu_0 M_s^2 \sum_{i,j=1}^{N} \int_{\Omega_i} \mathbf{m}(\mathbf{r}_i) \cdot \int_{\partial\Omega_j} \mathbf{m}(\mathbf{r}'_j) \cdot \hat{\mathbf{n}}(\mathbf{r}')g(\mathbf{r} - \mathbf{r}')\mathrm{d}^2\mathbf{r}'\mathrm{d}^3\mathbf{r}$$

$$= -\frac{1}{8\pi}\mu_0 M_s^2 \sum_{i,j=1}^{N} \mathbf{m}(\mathbf{r}_i) \cdot \int_{\Omega_i} \int_{\partial\Omega_j} \mathbf{m}(\mathbf{r}'_j) \cdot \hat{\mathbf{n}}(\mathbf{r}') g(\mathbf{r}-\mathbf{r}') \mathrm{d}^2\mathbf{r}'\mathrm{d}^3\mathbf{r}$$

$$= -\frac{1}{8\pi}\mu_0 M_s^2 \sum_{i,j=1}^{N} \mathbf{m}^T(\mathbf{r}_i) \left[\int_{\Omega_i} \int_{\partial\Omega_j} g(\mathbf{r}-\mathbf{r}')\hat{\mathbf{n}}^T(\mathbf{r}')\mathrm{d}^2\mathbf{r}'\mathrm{d}^3\mathbf{r} \right] \mathbf{m}(\mathbf{r}_j).$$

3×3 interaction tensors between cells Ω_i and Ω_j can be defined as

$$N(\mathbf{r}_i, \mathbf{r}_j) = -\frac{1}{4\pi|\Omega_i|} \int_{\Omega_i} \int_{\partial\Omega_j} g(\mathbf{r}-\mathbf{r}')\hat{\mathbf{n}}^T(\mathbf{r}')\mathrm{d}^2\mathbf{r}'\mathrm{d}^3\mathbf{r}.$$

Then, the energy approximation reads

$$E_{\mathrm{demag}} \approx \frac{1}{2}\mu_0 M_s^2 \sum_{i,j=1}^{N} |\Omega_i|\mathbf{m}^T(\mathbf{r}_i) N(\mathbf{r}_i, \mathbf{r}_j)\mathbf{m}^T(\mathbf{r}_j),$$

and

$$E_{\mathrm{demag}\,i,j} = \frac{1}{2}\mu_0 M_s^2 |\Omega_i|\mathbf{m}^T(\mathbf{r}_i) N(\mathbf{r}_i, \mathbf{r}_j)\mathbf{m}(\mathbf{r}_j)$$

is the exact energy between two uniformly magnetized cells Ω_i and Ω_j. The *demagnetizing tensors* $N(\mathbf{r}_i, \mathbf{r}_j)$ only depend on the difference vectors $\mathbf{r}_i - \mathbf{r}_j$, what justifies the abbreviating notations

$$N(\mathbf{r}_i, \mathbf{r}_j) = N(\mathbf{r}_i - \mathbf{r}_j) = N_{i-j}.$$

The demagnetization field $\mathbf{H}_{\mathrm{demag}}$ can now be extracted from the energy expression. $\mathbf{H}_{\mathrm{demag}}$ can be computed in cell Ω_i as

$$\mathbf{H}_{\mathrm{demag}}(\mathbf{r}_i) = -\sum_{j=1}^{N} N_{i-j}\mathbf{m}_j \tag{8.8}$$

where $\mathbf{m}_j = \mathbf{m}(\mathbf{r}_j)$. Eq. (8.8) is a discrete convolution of the demagnetizing tensors and the spontaneous magnetization (cp. Def. 2.24). For the tensors N_{i-j} the following properties hold (see [116, 52]):

- $N_{i-j} \in \mathrm{symm}(\mathbb{R}^{3\times3})$

- $\mathrm{tr}(N_{i-j}) = \begin{cases} 0 & \text{if } i \neq j \\ 1 & \text{if } i = j \end{cases}$

- The entries of N_{i-j} can be analytically calculated using integral expressions for the charges on parallel and orthogonal sides of interacting parallel-epipeds. The exact solutions are given by Newell et al. in [116] or Miltat and Donahue [52].

To calculate $\mathbf{H}_{\mathrm{demag}}$ for a finite extended specimen, FFT methods are applied. The demagnetization tensors N_{i-j} are initially calculated for all possible difference vectors $\mathbf{r}_i - \mathbf{r}_j$ in the computation domain. The exact solution formulae for the tensor components, based on an implementation provided by OOMMF[1], are used. The Fourier transforms of the N_{i-j} are stored initially, as their values do not change over time. To calculate the demagnetization field in time step n, the Fourier transform of the actual values of $\mathbf{m}$ are computed, and the discrete version of the convolution theorem Thm. 2.11 is applied to compute the convolution Eq. (8.8). This procedure is more efficient than a direct evaluation of Eq. (8.8). For the computation of the Fourier transforms, the free software library FFTW[2] is used.

8.2.2 3D periodic extended specimen

The method to calculate the demagnetization field differs significantly if the specimen under considerations is supposed to be an RVE cut out of a larger material sample, i.e. if Ω is periodic in all three spatial directions. The magnetization field $\mathbf{m}$ is assumed to be periodic, and Ω as a unit of repetition. Under this assumption, the solution for $\mathbf{H}_{\mathrm{demag}}$ can be derived directly from Eq. (4.3), and can be directly solved in Fourier space. For a function $f : \mathbb{C} \to \mathbb{C}$, by $\hat{f}$ the Fourier transform of f is denoted (see Def. 2.24). The Fourier transform of the Laplace-type equation Eq. (4.3) reads

$$(ik_1)^2\hat{\psi}(\mathbf{k}) + (ik_2)^2\hat{\psi}(\mathbf{k}) + (ik_3)^2\hat{\psi}(\mathbf{k}) =$$
$$M_s\left((ik_1)\hat{m}_1(\mathbf{k}) + (ik_2)\hat{m}_2(\mathbf{k}) + (ik_3)\hat{m}_3(\mathbf{k})\right),$$

---

[1]The Object Oriented MicroMagnetic Framework project is seated at ITL/NIST. The software OOMMF provides a framework to carry out micromagnetic simulations. The web presence can be accessed at `http://math.nist.gov/oommf/`
[2]Fastest Fourier Transform in the West: `http://www.fftw.org/`

and so the Fourier transform of the potential ψ is given by

$$\hat{\psi}(\mathbf{k}) = -iM_s \frac{k_1\hat{m}_1(\mathbf{k}) + k_2\hat{m}_2(\mathbf{k}) + k_3\hat{m}_3(\mathbf{k})}{(k_1^2 + k_2^2 + k_3^2)}, \qquad (8.9)$$

where $\mathbf{k}$ is the wave vector in Fourier space and i the imaginary unit in $\mathbb{C}$ (cp. Ex. 2.1). Using the equality $\mathbf{H}_{\text{demag}} = -\nabla\psi$ and applying the derivation theorem Thm. 2.11 to Eq. (8.9), the demagnetization field becomes

$$\mathbf{H}_{\text{demag}}(\mathbf{r}) = \mathcal{F}^{-1}\left(M_s \frac{k_1\hat{m}_1(\mathbf{k}) + k_2\hat{m}_2(\mathbf{k}) + k_3\hat{m}_3(\mathbf{k})}{\mathbf{k}^2}\mathbf{k} \right).$$

At $\mathbf{k} = \mathbf{0}$, the equation is not well defined. An idea of Zhang and Chen [88] uses the decomposition of the magnetization in analogy to the decomposition of strain (cp. the book of Khachaturyan [110] and Eq. (7.9)). The spontaneous magnetization $\mathbf{m}$ is separated into a homogeneous Part $\bar{\mathbf{m}}$ and a heterogeneous part $\delta\mathbf{m}$ satisfying

$$\bar{\mathbf{m}} \equiv \text{const} \qquad \text{and} \qquad \int_\Omega \delta\mathbf{m} = \mathbf{0},$$

such that

$$\mathbf{m}(\mathbf{r}) = \bar{\mathbf{m}} + \delta\mathbf{m}(\mathbf{r}).$$

Then

$$\hat{\mathbf{m}}(\mathbf{k}) = \hat{\bar{\mathbf{m}}} + \widehat{\delta\mathbf{m}}(\mathbf{k}).$$

With this, the demagnetization field $\mathbf{H}_{\text{demag}}$ in Fourier space becomes

$$\widehat{\mathbf{H}_{\text{demag}}}(\mathbf{k}) = \begin{cases} \widehat{\overline{\mathbf{H}}_{\text{demag}}} & \text{if } \mathbf{k} = \mathbf{0} \\ M_s \dfrac{k_1\widehat{\delta m_1}(\mathbf{k}) + k_2\widehat{\delta m_2}(\mathbf{k}) + k_3\widehat{\delta m_3}(\mathbf{k})}{\mathbf{k}^2}\mathbf{k} & \text{if } \mathbf{k} \neq \mathbf{0} \end{cases}. \qquad (8.10)$$

The quantity $\overline{\mathbf{H}}_{\text{demag}}$ is calculated as $\overline{\mathbf{H}}_{\text{demag}} = N\bar{\mathbf{m}}$, where N is the demagnetization tensor of the specimen, determined by its shape, and $\bar{\mathbf{m}} = \frac{1}{|\Omega|}\int_\Omega \mathbf{m}(\mathbf{r})\,d\mathbf{r}$. The complete field calculation is then given by

$$\mathbf{H}_{\text{demag}}(\mathbf{r}) = \mathcal{F}^{-1}\left(M_s \widehat{\mathbf{H}_{\text{demag}}}(\mathbf{k}) \right) \qquad (8.11)$$

For some few shapes, the demagnetization tensor N is known (cp. [44, 46]), for example for a general ellipsoid with eigenvalues a, b, c and $a+b+c = 1$

$$N_{\text{ellipsoid}} = \begin{pmatrix} a & 0 & 0 \\ 0 & b & 0 \\ 0 & 0 & c \end{pmatrix}.$$

In the case of a sphere, N becomes

$$N_{\text{sphere}} = \frac{1}{3}\mathbf{I}.$$

The implementation of Eq. (8.11) is realized efficiently by using the FFT methods the library FFTW provides. In each time step, the Fourier transform of $\mathbf{m}$ has to be calculated, as well as the average magnetization $\bar{\mathbf{m}}$. The shape tensor N has to be known and fixed for each simulation. For the simulation in this work, where an RVE is assumed, the assumption that the specimen is embedded in a sphere is made, so that $N = N_{\text{sphere}}$.

8.3 Verification of the implementation of micromagnetic equations

The numerical procedures that are described in the last sections are implemented and integrated into the Pace3D software environment. In order to apply the FFT techniques, routines of the FFTW are used. These have the advantage that the number of grid-points in each direction does not need to be a multiple of two (as most other libraries demand). In addition, the routines of FFTW are parallelized by MPI. The Laplace operator occurring in the exchange field $\Delta\mathbf{m}$ (see Eq. (4.11)) can either be discretized directly using central differences or, as it contains spatial derivatives, solved by application of Thm. 2.11 in Fourier space. Tests comparing both implementations did not show significant differences in the simulation results. The same applies to higher order finite differences implementations of $\Delta\mathbf{m}$.

To verify the implementation and to compare the numerical results of micromagnetic simulations with results of other scientific groups, two of the well accepted μMAG standard problems of the Micromagnetic Modeling Activity Group at NIST[3] have been simulated: A hysteresis loop in a permalloy specimen, and a simulation of a dynamic pulse on a permalloy-like thin film.

8.3.1 μMAG standard problem #1

The μMAG standard problem #1 describes the micromagnetic simulations in a rectangular permalloy specimen with dimension $2\mu m \times 1\mu m$ and

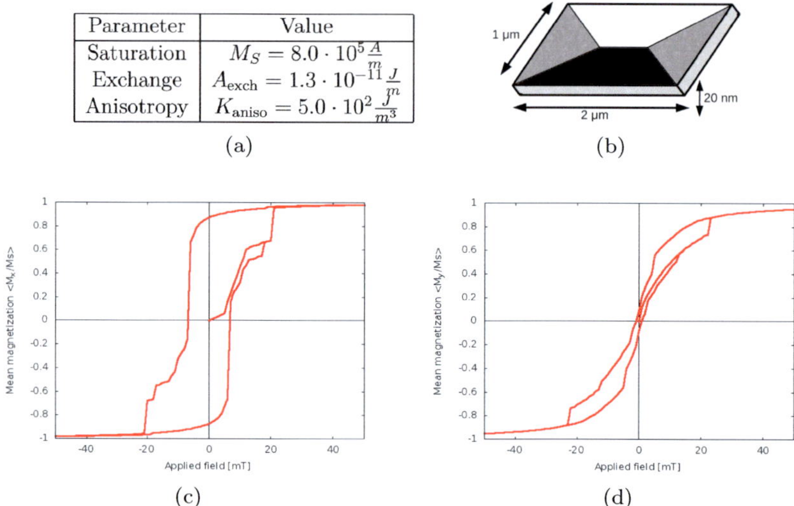

Figure 8.1: **(a)** Micromagnetic parameters and **(b)** geometry and dimensions for the μMAG standard problem #1. The experiments specified in the μMAG standard problem #1 were carried out numerically. **(c)** shows the numerical solution of the magnetization process in the direction of the easy axis (long edge), **(d)** the numerical solution in the direction of the hard axis (short edge). The results compare well with other results published on the μMAG homepage that are not shown here.

---

[3]`http://www.ctcms.nist.gov/~rdm/mumag.org.html`.

$20nm$ thickness. Magnetization vs. external field hysteresis loops shall be recorded. The specified parameters for the magnetic exchange, the uniaxial magnetocrystalline anisotropy and the saturation magnetization are $A_{\text{exch}} = 1.3 \cdot 10^{-11} \frac{J}{m}$, $K_{\text{aniso}} = 5.0 \cdot 10^2 \frac{J}{m^3}$ and $M_S = 8.0 \cdot 10^5 \frac{A}{m}$, the initial magnetization state is not specified. The direction of the unique easy axis is assumed to be parallel to the long edge of the rectangle. Figs. 8.1a and 8.1b show the parameters and a sketch of the setting. In the simulations the grid resolution was chosen to be $\Delta x = \Delta y = \Delta z = 20nm$ to coincide with the thickness of the specimen, so that the overall dimension of the calculation domain is $100 \times 50 \times 1$ grid points. The initial state of the magnetization consisted of randomly chosen magnetization vectors. The system was then let relaxed to gain a valid initial S-like state (cp. [52]). An external field parallel to the long edge was applied with increasing strength until the specimen was saturated, then the field was reversed. The same procedure was applied in direction of the short axis. The resulting hysteresis loops in the mean magnetization vs. applied field curves are shown in Figs. 8.1c and 8.1d. The results obtained compare quite well with results of other groups that are published on the μMAG homepage. These are not shown here, but can be accessed on the μMAG homepage (cp. footnote 3 on page 146).

8.3.2 μMAG standard problem #4

To analyze the time-evolution under the application of an external magnetic field, the μMAG standard problem #4 specifies a pulse experiment. A film of 3nm thickness, 500nm length and 125nm width is defined, exhibiting the same parameters as the permalloy rectangle from the μMAG Standard Problem #1 (see Fig. 8.1a), but showing no magnetic anisotropy (i.e. $K_{\text{aniso}} = 0\frac{J}{m^3}$). Initially, the film is in an 'S-state'. Figs. 8.2a and 8.2b show the geometry and the initial magnetization configuration. Then, two experiments with two different external fields are applied to the same initial state of the film:

- $\mu_0\mathbf{H}_1 = (-24.6, 4.3, 0.0)^T mT$, which is a field of approximately 25 mT, with a direction of $170°$ counterclockwise from the positive long axis of the parallel-epiped

- $\mu_0\mathbf{H}_2 = (-35.5, -6.3, 0.0)^T mT$, which is a field of approximately 36 mT, with a direction of 190° counterclockwise from the positive long axis of the parallel-epiped

Each field is applied until saturation is reached. The average magnetization in the **x**-, **y**- and **z**-direction vs. time is tracked. The domain used for the simulations has a dimension of $100 \times 25 \times 1$ grid points with uniform grid spacing of $\Delta x = 5$nm (so, the specimen is slightly thicker in the **z**-dimension as specified by the μMAG group). The numerical results are compared to the results Berkov et al. obtained by using the software

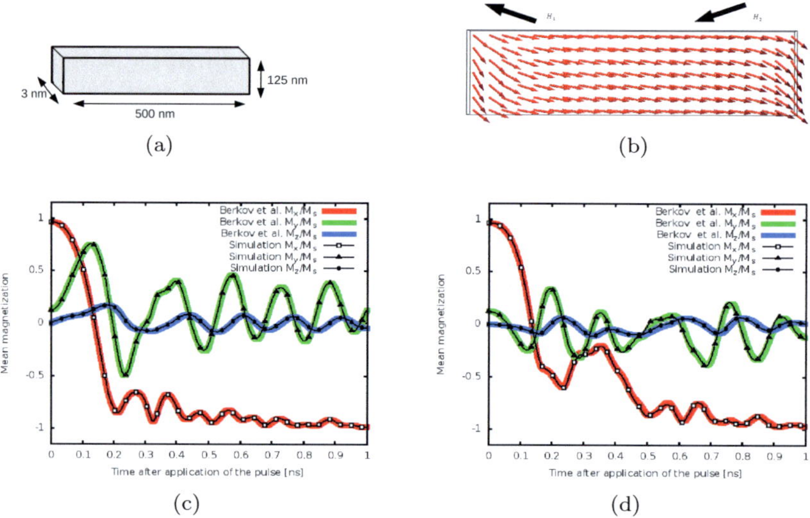

Figure 8.2: **(a)** Sketch of the setup for the standard problem # 4 and **(b)** the initial magnetization state. The fields $\mathbf{H}_1$ and $\mathbf{H}_2$ are applied in the indicated direction having approximately 25 mT and 36 mT, respectively. The numerical results are compared to the results of Berkov et al. Shown are the average magnetization vs. time curves for the external applied fields of **(c)** $\mu_0\mathbf{H}_1 = (-24.6, 4.3, 0.0)^T$ mT and **(d)** $\mu_0\mathbf{H}_2 = (-35.5, -6.3, 0.0)^T$ mT. As can be seen, the results compare very good.

MicroMagus[4] and that are published on the μMAG homepage. As can be seen in Figs. 8.2c and 8.2d, the results compare very well.

---

[4]`http://micromagus.de/`

Part IV

Application and Outlook

9 A phase-field model for polycrystalline thin film growth

The phase-field model of Sec. 6.2 of Chap. 6 is applied to model competitive grain growth on thin films, using the example of MFI zeolite-like coffin shaped crystallites as a model system. The presented results are published as an article in the Journal of Crystal Growth [3], and this chapter closely follows this article in text and structure. All figures presented in this chapter are taken from this article. The phase-field model used for the analysis shows as a first application the simulation of grains growing into a liquid and does not account for elastic or magnetic free energy contributions. The driving forces between the crystallites growing into the liquid are considered constant. So, the bulk free energy term f in Eq. (6.5) becomes constant for each phase. Values only differ between the solid and liquid phases (but not between different solid phases that represent the crystallites). In the following discussion, the bulk free energy density of the liquid phase will be denoted by f_{liquid}, the bulk free energy density of the solid phases by f_{solid}.

9.1 Introduction

Polycrystalline thin films are of high importance as catalytic active supports, especially for many reactions of technical interest [117]. This chapter focuses on modeling the growth evolution of zeolites on thin films that are widely used in conditions with high fluid flow rates or strong thermo-mechanical load, as in the catalytic cracking of petroleum hydrocarbons. The atomic microstructure of zeolites is characterized by a high amount of internal pores of about 5 to 10 nm size, important for their use as molecular sieves to separate gas mixtures of hydrogen and hydrocarbons at high temperatures. The crystalline structure of zeolites is characterized by interlinked silica and alumina tetrahedra, where the aluminum sites provide the catalytic active centers. Zeolite films are grown on supports in an autoclave, a reaction vessel which allows for high temperatures and pressures, from a hydrothermal solution. Their polycrystalline structure, influenced strongly by the seeding procedure and the support morphology, may give rise to a larger scale porosity by pinholes, domes or cracks created in between the different growing crystallites [118]. This porosity can ruin selectivity in the application as molecular sieve, so relatively thick membranes of about a few μm up to 50 μm must be grown to get a membrane free of pinholes and cracks [117]. On the other hand, to enable a large gas flux through the layer, its thickness should be as small as possible.

For the final morphology of the film, the nucleation stage is an important factor, where orientation and size distribution of the zeolite crystals have a major influence. Apart from direct growth of the silicate mineral on a support with usually random orientations of the nuclei, an effective route is the secondary synthesis, where seed crystals are deposited on top of the support. The seeds, often exhibiting a highly anisotropic shape, are grown in a first step by homogeneous nucleation from an amorphous silicate gel, then are cleaned and spread on the support in a colloidal solution, in the amount of one monolayer. This technique is commonly used to produce polycrystalline membranes of the MFI-type[1] (or ZSM-5 or silicalite-1), a model system in the study of zeolite growth on which is focused also in this study. Often, the evolution of a crystallographically preferred orientation is observed [119, 120], indicated by pole figures of the fully grown films.

Due to the anisotropic internal pore geometry related to the crystalline structure (see Fig. 9.1), the mass transport rate through a membrane is strongly anisotropic (that is dependent on the direction in space) and also depends on the orientation of the growing crystallites. For MFI-type zeolites, diffusivities can be more than four times larger for mass transport perpendicular to the c-axis than parallel to it [121]. On the other hand, crystal interfaces represent an even stronger diffusion barrier for the flux of permeating molecules. Hence, the goal of process optimization by simulation is the improvement of MFI membranes regarding flux

---

[1] The abbreviation is derived from the company name <u>M</u>obile <u>F</u>ive.

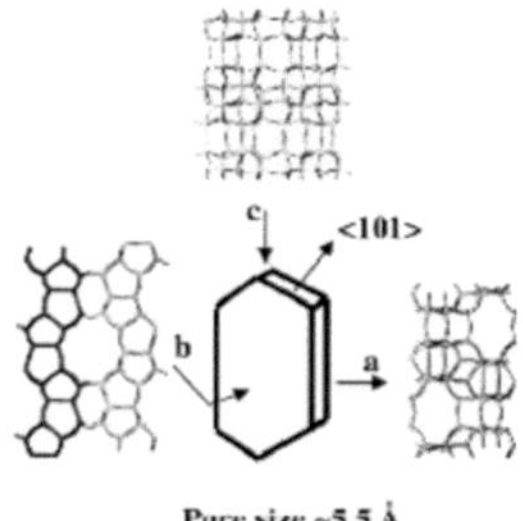

Figure 9.1: Internal pore geometry of zeolite crystallites. The figure is taken from the work by I. Díaz et al. [122]

rate, selectivity and mechanical (thermal) stability. To define a general objective, the goal is the production of closed thin films with a preferred orientation, isotropic in-plane texture and without secondary porosity or enhanced roughness.

To optimize and assess possible process modifications, simulations of microstructure evolution may be of great help. In every approach of modeling, phase transitions as well as grain growth, the treatment of the boundaries of homogeneous phases or grains is a crucial point. Moving boundary problems, as the growth of a crystal from a solution, require the application of special boundary conditions to account for the conservation of solute or heat in the process. Different from various front tracking approaches (see e.g. [123]), the phase-field model of Sec. 6.2 introduces an additional scalar parameter which varies continuously in space and time to describe the location of the interface. In the last decades a broad spectrum of phase-field models, including single and multiple parameter models, have been developed, mostly in the context of materials science (see [124] for a review). Many of the previous phase-field studies on polycrystalline growth in undercooled melts involved the introduction of an additional orientation parameter field with separated evolution dynamics to differentiate between the grain orientations, based on the work of Kobayashi et al. [125] and Warren et al. [126]. This approach was applied for diffusion coupled dendritic and spherulite growth [127] and in a study of growth competition for two silicon grains in a thermal gradient [128]. For the case of strong interfacial anisotropy, Eggleston et al. applied a regularization method which allowed for the description of crystal facets and missing orientations [129]. The crystal growth process in 2D as a combination of capillary and kinetic effects has been studied by Yokoyama and Sekerka analytically [130] and by Uehara and Sekerka using a phase-field model, including the formation of facets [131].

9.2 Modeling of polycrystalline thin film growth

For the use as membranes in filters or catalytic reactors, zeolite films are typically grown on a mesoporous metal or ceramic support. Here, a flat and smooth support of zeolite crystals grown in a preceding step is assumed, as it is the case in the seeding supported crystallization route. For this study, the orientation distribution of the seeds is assumed to be uniform, where each seed has its unique orientation that is unchanged over time. The growth of MFI zeolite films is a well examined model system for hydrothermal zeolite growth. In experimental studies, continuous growth conditions were achieved at least for a major period, which is reflected in a linear increase of film thickness in time (see e.g. [120]). Typical technical routes use additional *structure-directing agents* (SDAs) in the hydrothermal solution, which adsorb on the crystal faces and govern the attachment and integration of subcolloidal silicate particles from solution [122]. Hence, under these conditions a transport limitation by solute diffusion in the hydrothermal solution plays a minor role and nucleation of new crystallites can be excluded. Due to these findings and for the sake of simplicity, the driving force for crystallization is chosen to be constant.

9.2.1 Anisotropy function and single crystal shape

In general, the shape of a growing crystal results from the effect of both surface energy anisotropy and kinetics, the latter setting limits to the attachment of material on the growing interface and the long range transport through the liquid [132]. To examine the influence on the polycrystalline growth, faceted crystals which are formed by pure surface energy anisotropy (hence exhibiting their *Wulff shape*) and crystals formed by pure kinetic anisotropy (exhibiting their *kinetic Wulff shape*, see [132]) are studied. The former case corresponds to a slow growth near equilibrium, the latter to a fast growth mode. The two cases represent the possible extremes, whereas in reality both mechanisms could be important. As there is no precise information on the equilibrium Wulff shape, it is necessary to check whether in the simulation of polycrystalline growth an influence of anisotropy in the surface free energy on the force balance at triple junctions is present. This is a statement of the Gibbs-Thomson-Herring equation and found in previous theoretical as well as numerical studies [85, 133]. Hence, here the formation of the known zeolite crystal shapes formed by pure surface energy anisotropy as well as by pure kinetic anisotropy is studied, keeping in mind that in physical reality both effects will interfere.

In experiments, the dominant growth shape of ZSM-5 (MFI) zeolites using TPA^2 as an SDA under moderate conditions is the hexagonal prismatic or coffin shape, see Fig. 9.3b taken from reference [122]. Due to the orthorhombic symmetry of the ZSM-5 zeolite (an analogue of the natural mineral mutinaite [134]), the crystal shape exhibits three mirror planes perpendicular to the *a*-, *b*- and *c*-axis directions (Fig. 9.2b). Its exposed facets are (1 0 0), (0 1 0) and (1 0 1). According to the dimensions of the unit cell [135], the ⟨1 0 1⟩ direction is tilted by 33.7° with respect to the *c*-axis, not by 30°, as for the ideal hexagon. Nevertheless, despite of the small introduced error, the crystal shape is modeled by 120° internal edge angles for the (0 1 0) facets as in an ideal hexagon. Furthermore, within this study recent findings are neglected that state that the prismatic crystal could be a composition of six different twin components [136]. Also, growth twins appearing at the (0 1 0) faces under certain conditions (and rotated by 90° with respect to the parent crystal *c*-axis) are not considered. A minor influence of the growth twins under the studied growth conditions is expected in this study.

In accordance with a previous simulation study [137] and experimental results [138], crystals with typical aspect ratios were chosen, exhibiting a tip to tip extension in *c*- vs. *a*-direction^3 of 2 : 1 and 4 : 1. The extension

---

^{2}tetrapropylammonium
^{3}length to width, or *c* : *a* aspect ratio

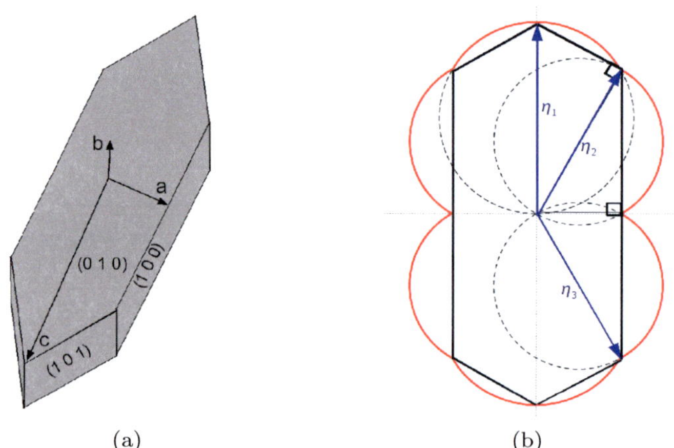

Figure 9.2: **(a)** A typical 3D coffin shape with crystal axes indicated. The aspect ratio is defined here as ratio between the extension along the c-axis vs. the a-axis. **(b)** Plot of the surface energy anisotropy function $a_{\alpha\beta}(\theta)$ (red) as a function of the polar angle for a 2D hexagonal prismatic crystal with $2:1$ aspect ratio. A sketch of the Wulff construction for two facets including three vertex vectors $\vec{\eta}_i$ is shown.

along the b-direction was always fixed to half of the extension along the a-direction in the 3D simulations, giving the usual habit of an elongated flat hexagonal platelet as shown in Fig. 9.3b. In the model, crystal anisotropy is a function of the gradients of the phase fields ϕ_α defined in a fixed reference coordinate system Eq. (6.7). To describe various orientations, the gradients are properly transformed using the three Euler rotation matrices for the different orientations of the grains (cp. Appendix A.1). The non-rotated standard crystal shape is described via an aspect ratio along the three axes as $c:b:a = 4:0.5:1$, fixed in the setting as the **x**-, **y**- and **z**-axes of the reference coordinate system.

Surface energy anisotropy For the case of surface energy anisotropy in 3D, 12 vertex vectors of the Wulff shape used in the anisotropy function Eq. (6.8) are defined for the $1 : 0.5 : 2$ crystal morphologies as

$$\vec{\eta}_{1,\cdots,4} = \begin{pmatrix} \pm 1 \\ \pm 0.25 \\ 0 \end{pmatrix} \quad \text{and} \quad \vec{\eta}_{5,\ldots,12} = \begin{pmatrix} \pm(1 - \frac{1}{\sqrt{3}}) \\ \pm 0.25 \\ \pm 0.5 \end{pmatrix}, \quad (9.1)$$

and for the $1 : 0.5 : 4$ shape as

$$\vec{\eta}_{1,\cdots,4} = \begin{pmatrix} \pm 1 \\ \pm 0.25 \\ 0 \end{pmatrix} \quad \text{and} \quad \vec{\eta}_{5,\ldots,12} = \begin{pmatrix} \pm(1 - \frac{1}{4\sqrt{3}}) \\ \pm 0.25 \\ \pm 0.25 \end{pmatrix}. \quad (9.2)$$

The facet energies of $\sqrt{3}/2\,\gamma_{\alpha\beta}$ for the $(1\ 0\ 1)$, $1/4\,\gamma_{\alpha\beta}$ for the $(1\ 0\ 0)$ and $1/8\,\gamma_{\alpha\beta}$ for the $(0\ 1\ 0)$ facets can be derived from the Wulff construction. For the 2D simulations, the respective **y**-components in all vertex vectors in Eqs. (9.1), (9.2) and (9.3) are left out. In Fig. 9.2b the corresponding plot of the anisotropy function for the $2 : 1$ shape as a function of the polar angle is depicted, showing also a sketch of the Wulff construction from which the equilibrium crystal shape (bold solid line) is obtained.

Kinetic anisotropy In case of pure kinetic anisotropy, the interface evolution is modified by the orientation dependent coefficient τ (Eqs. (6.10) and (6.12)), which modulates the normal interface velocity, whereas the interface tension $\gamma_{\alpha\beta}$ is constant for all evolving facets. To reproduce the coffin shape, the easiest approach is to create first an equiaxed hexagonal platelet with the anisotropy function Eq. (6.8) and the following vertex vectors

$$\vec{\eta}_{1\cdots 4} = \begin{pmatrix} \pm 1 \\ \pm 0.25 \\ 0 \end{pmatrix} \quad \vec{\eta}_{5\cdots 12} = \begin{pmatrix} \pm 0.5 \\ \pm 0.25 \\ \pm 0.5\sqrt{3} \end{pmatrix}. \quad (9.3)$$

Second, to adjust the desired $c : a$ aspect ratio, this shape is modulated with an elliptical anisotropy with the **x** direction as semimajor axis, given by

$$a^{ellips}(\phi, \nabla\phi) = \left(1 - \frac{\delta_y(q_{\alpha\beta})_y^2 - \delta_z(q_{\alpha\beta})_z^2}{|\vec{q}_{\alpha\beta}|^2}\right). \quad (9.4)$$

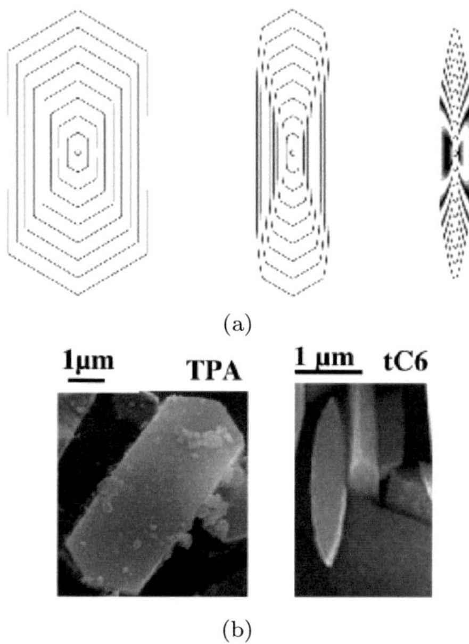

Figure 9.3: **(a)** Contour lines of the phase-field parameter ($\phi = 0.5$) for three different values of the elliptical parameter $\delta = 0.5, 0.25$ and 0.1 in the kinetic anisotropy corresponding to the $2:1$, $4:1$ and leaf-like shape (from left to right). The temporal spacing is 1000 time steps (simulation parameters are given in Chap. 9.4). **(b)** SEM images of typical coffin and leaf-like shaped crystals grown with SDAs TPA and tC6, reprinted in [3] with permission from [122], Fig. 1. Copyright 2004 American Chemical Society.

Hence, the **x** direction is the one of fastest growth. When taking the parameter $\delta_y = \delta_z = \delta$ in Eq. (9.4), the $a:c$ aspect ratio of the crystal will be changed without modifying the $a:b$ ratio. The final function for the kinetic anisotropy takes the form

$$a_{\alpha\beta}^{kin}(\phi, \nabla\phi) = a_{\alpha\beta}(\phi, \nabla\phi)^{-4}\, a^{ellips}(\phi, \nabla\phi)^{-1}, \qquad (9.5)$$

where the anisotropy functions in Eq. (9.5) appear with negative powers as the inverse coefficient τ^{-1} in Eq. (6.10) is proportional to the interface normal velocity, which is given by $v_n = \partial_t \phi_\alpha / |\nabla \phi_\alpha|$. The forth power in the hexagonal anisotropy function is necessary to prevent the appearance of rounded corners, as the corresponding cusps in the γ-plot become deeper (and sharper). It is necessary to mention that this also reduces the solid-liquid mobility compared to simulations with the same driving force and surface energy anisotropy, as the facet velocity then scales with $(\sqrt{3}/2)^4 \approx 0.56$. The standard $4:1$ crystal shape in the simulations can be produced with $\delta = 0.75$, the $2:1$ shape with $\delta = 0.5$.

In experimental work the growth rate in the a- and b-crystal directions has been found to depend on the specific choice of an SDA, what leads to a change of the aspect ratio in the cross section perpendicular to the c-axis (long direction). An interesting feature concerns the value of the ellipsoidal parameter δ: when decreasing it below a value of 0.1, the $(1\,0\,1)$ facets disappear completely, and become slightly curved, giving rise to a leaf-like shape (see Fig.9.3a). This shape is observed in ZSM-5 growth moderated by the SDA tC6 [122], which seems to suppress growth selectively along the a-direction.

9.2.2 Treatment of polycrystalline orientations

In the modeling problem, interfaces between grains and the liquid phase as well as between different oriented grains are encountered. In the latter case, the interfacial free energy $\gamma(\vec{q}_{\alpha\beta}/|\vec{q}_{\alpha\beta}|, S_{\alpha\beta})$ would be generally a function of grain boundary inclination, in this model given by the norm of the generalized gradient vectors $\vec{q}_{\alpha\beta}$, and the misorientation $S_{\alpha\beta}$ between different grains. For each grain, the orientation is given by the triple of Euler angles $(\varphi_1^\alpha, \varphi_2^\alpha, \varphi_3^\alpha)$, interpreted as rotations around the fixed Cartesian axes in the frame of reference (cp. Appendix A.1). The misorientation matrix between two grains α and β is generated from the corresponding rotation matrices as $S_{\alpha\beta} = R^\beta (R^\alpha)^{-1}$ and used to transform the interface normals, important to treat a possible interface energy dependency. More important, corresponding grain boundary energies used as constants $\gamma_{\alpha\beta}$ in the model function of Eq. (6.7) would be calculated as function of misorientation according to an appropriate relationship (e.g. Read-Shockley). Contrary to phase-field models applying an orientation parameter field (see e.g. [126]) this approach allows for grains having identical orientation and definite boundary energy. In metals, this is encountered for the case of antiphase domain boundaries, and is observed in zeolites in form of purely translational grain boundaries [139]. This is a quite common phenomenon due to the large unit cell. In a TEM microscopy study of grain boundaries, de Gruyter et al. [139] also state in their conclusion that grain growth in zeolites does not follow energy minimization. Therefore, in this study all grain boundaries are treated to be isotropic. Within the formulation of the model, this corresponds to setting all parameters $\gamma_{\alpha\beta}$ to an identical value and no inclination dependency occurs, i.e. $a_{\alpha\beta} = 1$, cp. Eq. (6.8). This is a compromise due to the limited knowledge of grain boundary energies in zeolites, and does not represent a general drawback of the model.

For the initialization of the simulations, a sound distribution of the orientations of the seed crystals is necessary, each represented by a single order parameter. As the computational resources available are limited, an optimal equidistribution of orientations is preferred. For the analysis of the selection mechanism the number of generated orientations should be as high as as possible, which can be reduced exploiting the orthorhombic symmetry of the coffin shapes under consideration. For simulations in 2D one can restrict to equidistantly divide the interval $[-90°, 90°[$ in steps of $1°$. In 3D, the problem is more complicated. Because equidistant point distributions on the unit sphere $\mathbb{S}^2$ are hard to compute, the following approximative approach is taken. An icosahedron, which is a Platonic solid of type $\{3, 5\}$ (with 12 points, 30 edges and 20 equilateral triangular facets, cp. [25]), is inscribed into the unit sphere $\mathbb{S}^2$. In analogy to the Sierpiński tessellation of a triangle in fractal geometry, iteratively for each triangular facet new points on $\mathbb{S}^2$ are generated as follows:

1. Compute the three midpoints of the triangles' sides.

2. Project these points onto the unit sphere and add them as new points.

3. Use the new points and the triangles' old vertices to create four new equilateral triangular facets.

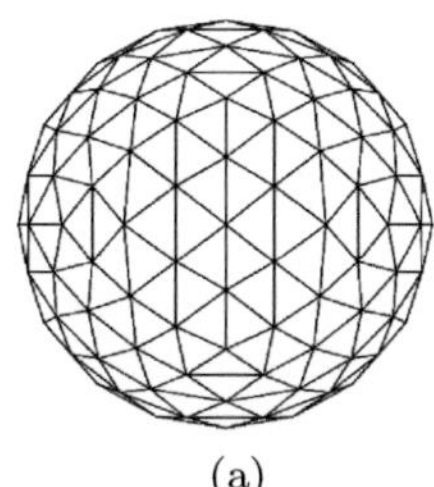
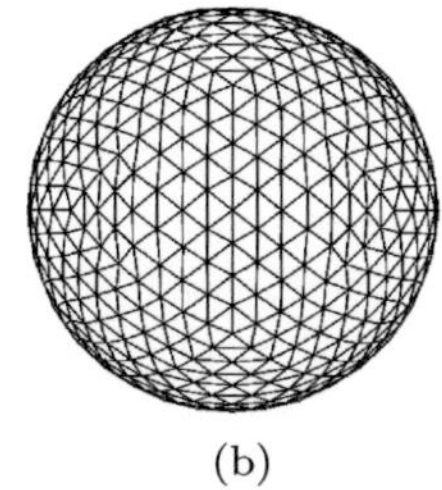
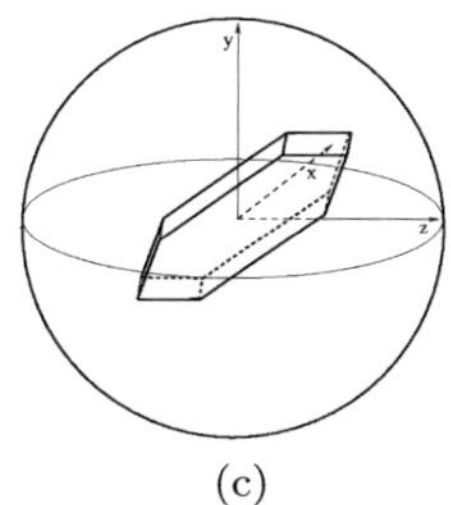

(a)   (b)   (c)

Figure 9.4: The polyhedron after **(a)** two and **(b)** three tiling steps of an initial icosahedron, whose vertex points lie on the unit sphere $\mathbb{S}^2$. **(c)** shows an oriented zeolite crystal in $\mathbb{S}^2$.

Because all points are on the unit sphere, the resulting figures are always convex, and the number of points after $i \in \mathbb{N}$ tiling steps is given as

$$P_i = 10 \cdot 4^i + 2.$$

The obtained point distribution is a sufficiently good approximation to a uniform point distribution on the unit sphere (cp. Figs. 9.4a and 9.4b), although poles exist around the original twelve vertices of the icosahedron. The generated points may serve as directions of growth for the crystal seeds building a thin film, whereas, again due to the symmetry of the assumed shapes and growth direction, consideration is restricted to points generated on the positive half sphere $\mathbb{S}^2 \cap \{(x, y, z) \in \mathbb{R}^3 | x \geq 0\}$. As for the case of secondary growth at $175°$ C studied in [120], it is assumed that the orientations of the starting crystals are uniformly distributed and that all seeds have the same size. Initially, all 2D simulations start with an already intergrown flat film, i.e. without spaces between the crystals. The 'seeding algorithm' in 3D for equisized and equishaped seeds works as follows:

1. Tile the y-z-plane into equisized squares.

2. Place a seed in each square as an ellipsoidal cap not touching any neighbour.

3. From the point distribution choose uniquely a direction and set it as the growth direction for exactly one seed.

4. Rotate each seed about a random angle drawn from $[-90°, 90°[$ around its c-axis, the preferred growth direction.

The last step of the algorithm is necessary to account for the orthorhombic symmetry of the crystal. Therefore, when adjusting the orientation in 3D, there are three independent angular degrees of freedom. To generate the oriented crystals initially, rotations around the internal c-axis (equivalent to the $\mathbf{x}$ axis of the reference frame) and two successive rotations around the perpendicular $\mathbf{y}$ and $\mathbf{z}$ axes are carried out. Fig. 9.4c illustrates an obliquely oriented crystal in the fixed coordinate frame. In 2D as well as in 3D, directions with small deviation from the substrate normal are referred to as 'normal' or 'straight' directions, others are referred to as 'oblique' directions.

9.3 Setup and simulation parameters

All simulations were performed with dimensionless parameters by choosing dimensional scale values (indicated by the subscript zero), e.g. a length scale d_0 for the dimensionless spatial coordinate $x = \tilde{x}/d_0$. The scale for the free energy density is related to the interface tension $f_0 = \gamma_0/d_0$ and the time scale to the kinetic coefficient τ in the model by $t_0 = \tau_0 d_0^2/\gamma_0$ (see [140] for details). All relevant scale parameters used in the simulations are listed in Tab. 9.1 together with their dimensional values.

length $[m]$	time $[s]$	interface tension $[J/m^2]$	energy density $[J/m^3]$	kinetic coeff. $[Js/m^4]$
$d_0 = L/N_x$	t_0	γ_0	$f_0 = \gamma_0/d_0$	$\tau_0 = \gamma_0 t_0/d_0^2$
$5 \cdot 10^{-9}$	1.0	1.0	$2 \cdot 10^8$	$4 \cdot 10^{16}$

Table 9.1: List of all relevant dimensional scale values used in the simulations. The length scale is defined via the domain length perpendicular to the growth direction, $L = 18\,\mu m$ in the 2D simulation, and the respective number of grid points $N_x = 3600$.

Despite the numerical optimizations, the number of growing crystals still defines the computational complexity of the problem. In the simulations, one presumption is that growth starts from 'supercritical' seeds, which means that their size is large enough to balance the effect of bulk driving force and solid-liquid interface tension. To account for a sufficient numerical resolution of the diffuse interface, a dimensionless interface width parameter of $\xi = 8.0$ and a grid spacing of $\Delta x = 1.0$ are chosen. This leads to a diffuse interface resolved by at least 8 grid points, so that the initial seed size was chosen to be 20 grid points (in 2D and in 3D). This is equivalent to a diameter of $0.1\,\mu m$ applying the length scale given in Tab. 9.1, typical for the secondary zeolite growth process [120]. In 2D, a simulation box of 3600×1500 grid points (representing typical film dimensions of $18.5\,\mu m$) was used, in 3D a cube of 570 points ($2.85\,\mu m$). For the time update, a step width of $\Delta t = 0.25$ fulfills the stability criterion of the explicit algorithm.

Concerning the energetics in the model, data typical for zeolitic silica are used, namely surface enthalpy and transformation enthalpy. The choice of enthalpy values can be justified by the small contribution of surface entropy and the small volume differences in the zeolitic transformation [141]. An interface free energy parameter of $\gamma_{\alpha\beta} = \gamma = 0.1$, which matches the given surface enthalpy of $0.1\,J/m^2$ (Tab. 1 in [141]) is chosen for both solid-liquid and solid-solid interfaces in all simulations. The obstacle potential (Eq. (6.9)) was chosen with a higher order parameter $\gamma_{\alpha\beta\delta} = 1.5$. The formation enthalpy per mol SiO_2 of $10\,kJ/mol$ [141] is converted into an energy density of $3 \cdot 10^8\,J/m^3$ in dividing it by the molar volume of MFI zeolite of $34\,cm^3$. The driving force for crystallization, which is the difference of liquid and crystal bulk free energy densities appearing in Eq. (6.5), is in general a function of solute composition and temperature. For hydrothermal zeolite growth under high silica concentrations, Nikolakis and coworkers found an independence of the growth rate of single crystals from the silica content, the main constituent of solid zeolite [142], during a long period of the growth. The growth process was found to involve the attachment of nanoscale building units, the formation of a constant surface charge and energy activated steps for their incorporation. This is attributed here to a dominance of interface kinetic effects, and assumed the thermodynamic driving force to be constant, represented in the simulations by $f_{liquid} = 0$ and negative values of f_{solid} for different crystallization rates, which in the experiment would depend on the process temperature. This represents mainly the situation at the growth front. For isolated cavities and pores, the simplified assumption is not capable to describe the closing in physical situations.

As the process possibly involves one or more energetically activated steps, the formation enthalpy given can be seen as an upper limit for the driving force. The value $f_{solid} = -0.1$, about 10% of the maximum enthalpy change, is used throughout the simulations and compares well to a value of $0.514\,kJ/mol$ found in a calorimetric study [143].

For single MFI crystals, various $c : a$ shape ratios from 1.5 to values greater than 5 have been experimentally found, depending on the composition of the growth solution [138]. Analysis here is restricted to the typical case of a fast growth with fixed $4 : 1$ aspect ratio. A basic assumption applied in the following is that solid-liquid interface energies and kinetics in the polycrystal are the same as for the single crystal. The kinetic coefficient in Eq. (6.13) is defined for the solid-solid interfaces via $a_{ss}^{kin} = 1$ and for the solid-liquid interface by the function defined in Eq. 9.5 (s and l denote the phase fields of solid grains and liquid phase, respectively) with coefficients $\tau_{sl}^0 = 1.0$ and $\tau_{ss}^0 = 10.0$. Hence, the mobility of the solid-solid interfaces are reduced by a factor of ten, sufficient to prevent substantial grain coarsening behind the moving crystallization front. This assumption is well justified for hydrothermal growth temperatures of about 150° C, which would make grain boundary migration very improbable. Typical temperatures necessary to induce significant grain coarsening are about 800° C, which is necessary for zeolite powder sintering [144]. With the choice of isotropic grain boundaries, our simulation study includes a similar simplifying assumption as Chen et al. have used in the study of growth competition of two silicon grains during solidification [128], but extending this problem into a polycrystalline 3D setting.

For the present study one gets the scale value for τ in Tab. 9.1 inversely by comparing the growth rate in the simulation with experimental data, here from the article [120]. In principle, atomistic simulations could be used to get more specific values for this coefficient. To make simulation results comparable, the same model parameters have been used in the 2D as well as in the 3D simulations.

9.4 Simulation results

A basic aim of the study was to elucidate whether general results in thin film growth are reproduced by the phase-field model, and how the influence of two different interface properties (surface energy and kinetics) modify these findings. Also, the effect of different driving forces related to different crystallization temperatures in hydrothermal growth had to be examined. Simulations were carried out first in 2D to study the general growth dynamics and to optimize the parameter set regarding to experimental conditions. For selected data sets, large scale 3D simulations were carried out on a parallel computing cluster.

2D simulations The 2D simulations were carried out on a grid with a resolution of 3600×1500 grid points, with periodic boundary conditions in the directions orthogonal to the substrate normal and isolation conditions at the substrate and liquid boundary (cp. Sec. 7.2). The initial setting has a random distribution of 180 seeds, already intergrown as a flat film. Testing locally separated equisized spherical seeds gave no observable difference in the resulting film morphology, so that out a major influence of the seed shape can be ruled out.

A total of eight simulation settings were generated, differing only in the randomly generated initial assignment of orientations to seeds. Each simulation was then run in both of the growth modes, and the results were analyzed due to different criteria. In Fig. 9.5 several stages of a film growth simulation are shown. The chosen color scale for the grains reflects the deviation from the normal direction, and runs from blue $(-90°)$ to yellow $(+90°)$, so that grains growing in normal direction are colored in red. Interestingly, slight misorientations can grow steadily during the evolution, if a local accumulation of a tilt angle arises. This can be seen in the right half of Fig. 9.5 (d), where a bundle of narrow left and right tilted grains appears.

As the grains forming the film compete during growth, some grains are overgrown by others. This process was also studied in detail and is shown exemplarily in Fig. 9.6, where stronger misoriented grains are successively prevented from further growth by neighboring grains. Between the events of grain extinction, the liquid-solid-solid triple junctions move on straight

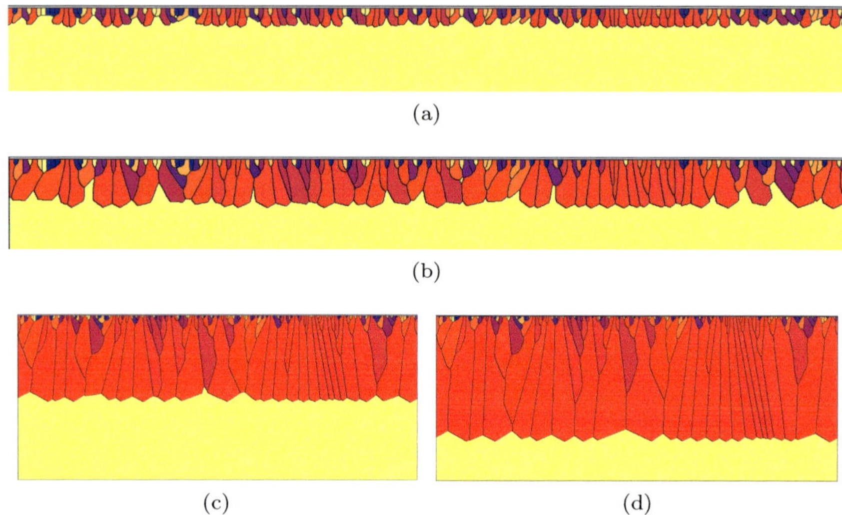

Figure 9.5: Two early stages **(a)** and **(b)** shown as a close-in on the front, after 2000 and 6000 time steps, and two late stages **(c)** and **(d)**, after 24000 and 35000 time steps, of competitive thin film growth with kinetic anisotropy. Dimensions of images (c) and (d) are scaled with 0.5 compared with (a) and (b). Growth competition and outgrowth of grains can be clearly observed.

lines. The analysis of the triple junction path shows that their direction is given by the average of the normals of the two (1 0 1)-facets in contact, $\theta_{ij}^{TJ} = \frac{1}{2}(\theta_i + \theta_j)$, where θ_i is the i-th grain orientation. Abrupt bending of the grain boundary occurs in two cases: Either one facet is completely consumed by the overgrowing grain and a facet with other inclination participates at the triple junction, or after a grain is completely overgrown and two previously unconnected grains come into contact.

A knowledge of the evolution of the orientation distribution is especially beneficial to interpret thin film diffraction experiments, which give either volume averaged results as for X-ray rocking curve measurements, or surface sensitive results as in RHEED (reflection high energy electron deflection) experiments. For a statistic evaluation of the simulation re-

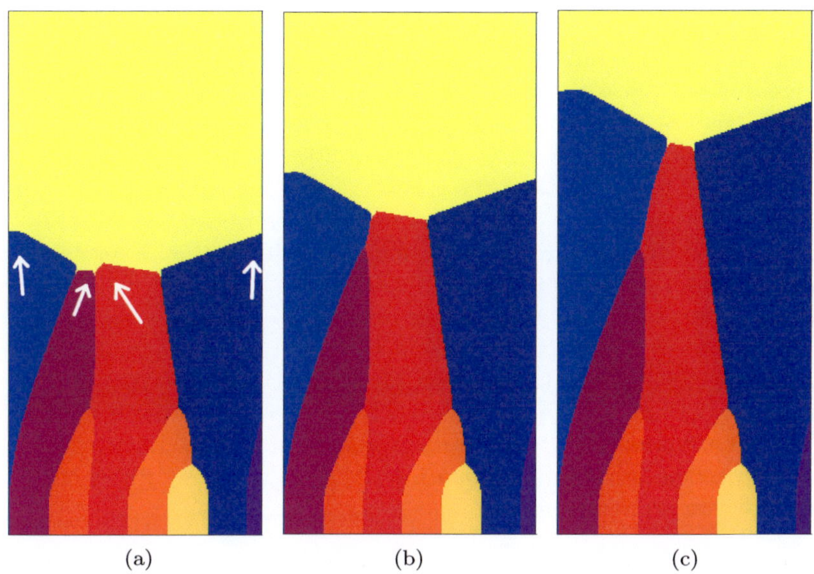

(a) (b) (c)

Figure 9.6: Overgrowth mechanism during growth competition: **(a)** Two grains with oblique orientations (purple and red), surrounded by two straight growing grains (blue). (0 0 1) directions of fastest growth are marked by white arrows. Co-evolution of light red grain and left neighbor continues, until the remaining left (1 0 1) facet of the oblique growing grain is fully consumed **(b)**, followed by a right-bending of the grain boundary **(c)**.

sults, the competing grains are classified due to their growth direction tilt from the normal direction. Both left and right tilted grains are collected into classes covering angular intervals of 10° from 0° to 90°. To account for the actual surface coverage of the orientations, the total lateral (in-plane) film width occupied by all grains within each orientation class was measured at several time steps and divided by the total film width. The evolution of the initially uniform distribution into a normal distribution is given in Fig. 9.7, characterized by the percentage of still growing grains. This plot can be interpreted as momentary surface occupied by grains of the respective orientation class. A monotonic increase in time in the spacial fraction of grains with orientation close to substrate normal for both growth modes is observed, where the grain distribution at each time is well represented by a normal distribution of the form $f(\theta) = \frac{2}{\sigma\sqrt{2\pi}} exp(-0.5(\theta/\sigma)^2)$. A quantitative measure for grain selection dynamics is the width of the actual orientation distribution vs. time. This is given as the standard deviation σ of the normal distribution in Fig. 9.7(b) as a function of the ratio of still growing grains. Especially the simulation with anisotropic surface energy shows a striking correlation between angular width and the number of growing grains $N(t)$, which gives a fit in the form of $\sigma(N(t)) = 58.3° N(t)/N_0 + 1.9°$, N_0 being the initial grain number. The results indicate, that the orientation distribution of the polycrystalline system relaxes quickly into a normal distribution, having a width σ which relates linearly to the number of competitors. This is accomplished by strictly eliminating the stronger tilted grains over the complete film growth process.

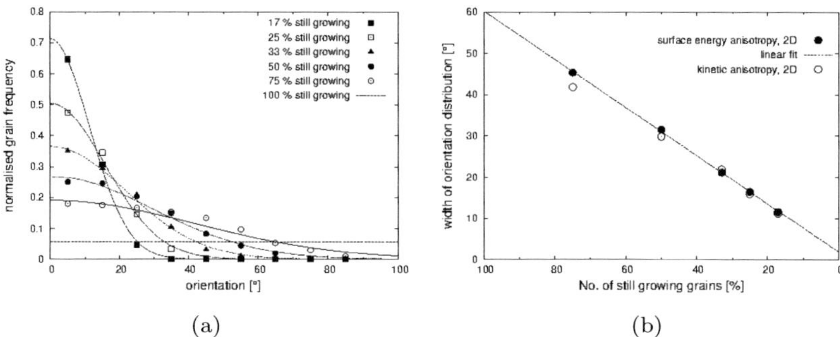

(a)   (b)

Figure 9.7: Dynamics of the selection mechanism due to deviation from substrate normal. **(a)** Distribution of orientation classes of still growing grains in intervals of $10°$, weighted with the occupied in-plane film width (symbols) and fitted Gaussians, for kinetic anisotropy. Results are averaged over eight simulations. **(b)** Development of the variances of the fitted normal distributions from (a).

Due to the specific formulation of kinetic anisotropy, the facet velocity is reduced by a factor of 0.56 compared to the surface anisotropy simulations,

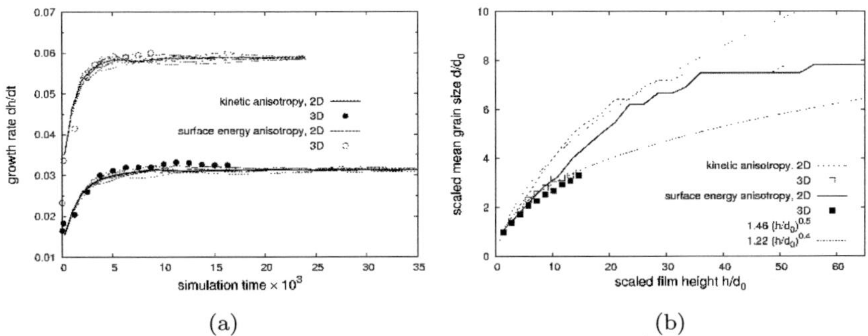

Figure 9.8: **(a)** Growth rate of the film front position for 2D (eight different simulations, light broken curves with average given as bold solid and dashed line) and 3D (open and closed circles). **(b)** Evolution of the mean in-plane grain size, scaled with the initial size d_0 vs. scaled film height. The expected power growth law in 2D and 3D is fitted for the case of kinetic anisotropy with the data for $0 - 35\,d_0$.

as mentioned in Sec. 9.2.1. In both cases the growth velocity of the whole film vs. time has roughly the characteristic of a simple exponential asymptotic to a constant value, cp. Fig. 9.8(a). To calculate growth speed, the average film height $\langle h(t) \rangle$ was computed for each recorded frame of the simulation as the total area (volume in 3D) of all solid grains divided by the box width L (lateral film surface A in 3D). Obviously, after an induction period of about $\Delta t = 5 \cdot 10^3$ to $8 \cdot 10^3$, the film height h increases linearly with time, in agreement with experimental results in [120].

In Fig. 9.8(b) the evolution of the mean grain size in the lateral film plane (= in-plane) d is given, scaled by the initial (seed) size d_0. In 2D, d is considered as grain diameter and is computed by division of the box width L by the actual grain number, in 3D it is computed from the mean grain area $\langle A \rangle = d^2 \pi / 4$. For the evolution a parabolic behavior $d(t) \propto h(t)^{1/2}$ in 2D and $d(t) \propto h(t)^{1/4}$ in 3D is expected from theoretical results [145, 146]. In the simulations, a film of height up to $50 - 70\,d_0$ was grown. To verify the growth exponent, a linear relation between film height and simulation time was assumed, which matched well except during a short initial transient (see Fig. 9.8(a)). The expected front dynamics fits well within a period of $0 - 35\,d_0$, illustrated with fine dotted curves in Fig. 9.8(b). After that, a clear retardation takes place for the 2D simulations. All remaining grain orientations are found within a small interval of about $10°$ around the normal direction, which drastically slows down the grain selection process. The stabilization of in-plane grain size motivates the need for further 3D simulations. It must be noted that the vertical box size of the 3D simulation was limited and the number of remaining grains too low to be statistically significant.

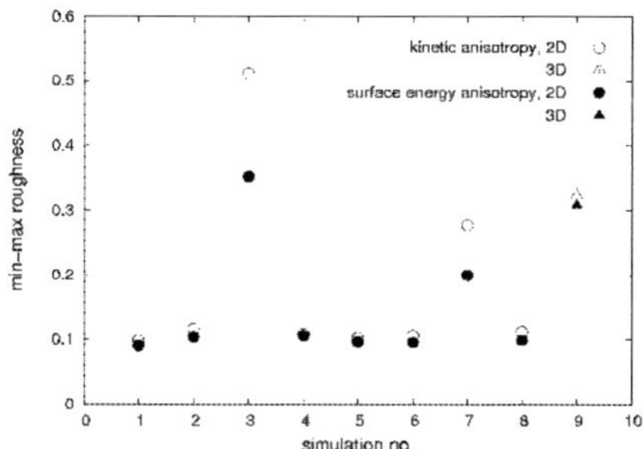

Figure 9.9: The roughness parameter $R_{min-max}$ (Eq. (9.6)) for each simulation for both growth modes. The dotted line indicates the value derived from pure geometric arguments, related to the maximum in-plane grain width and facet tilt angle.

According to the different grain orientations, there is a high local variation of growth speed in substrate normal direction. To characterize the resulting jaggedness of the film in regard of an application as membrane, a min-max roughness parameter is defined as

$$R_{min-max} = \frac{h_{\mathrm{max}} - h_{\mathrm{min}}}{\langle h \rangle},\qquad(9.6)$$

where h_{max} and h_{min} are the maximum and minimum film positions in substrate direction at the last step of the simulation, and $\langle h \rangle$ denotes the average front position. Different from the mean square roughness parameter, singular clefts or channels which would compromise the function as

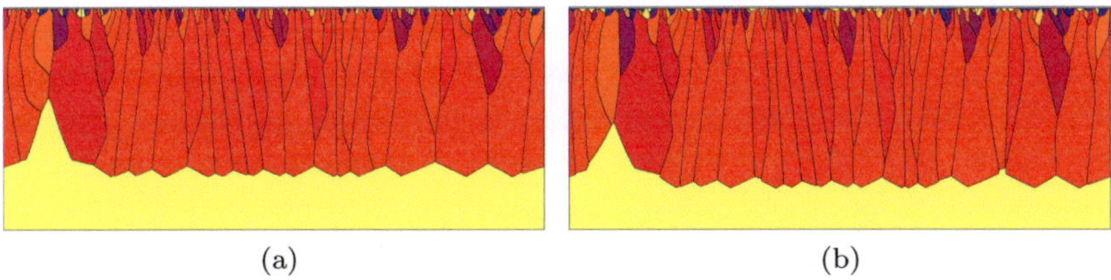

Figure 9.10: **(a)** Simulation of a rough film (kinetic anisotropy, after 35000 time steps) and **(b)** simulation from the same inital configuration (surface energy anisotropy, after 20000 time steps). The shape of the growth fronts and the surviving grains are highly comparable.

a membrane, determine its value. In Fig. 9.9 the value of the min-max roughness is plotted vs. the number of the simulation run for eight simulations in 2D and one in 3D, respectively. For most of the runs in 2D, $R_{min-max}$ has a narrow scatter around 10%, which is close the value of 9.2% derived from simple geometric considerations, taking into account the maximum occurring in-plane grain width, the tilt angle of 33.7° of the (1 0 1) growth facet with respect to the substrate normal, and assuming a closed film of grains with optimal growth direction at the final stage of the simulation.

Larger values up to about 50% stem from special initial orientation configurations, where several grains close to a substrate location are symmetrically left and right tilted, giving rise to the formation of V-shaped dips limited by slow growing (1 0 0) facets (simulations no. 3 and no. 7). Fig. 9.10 shows an example of this configuration. The high roughness of the 3D film given in Fig. 9.9 appears overestimated in comparison to the 2D films, as $R_{min-max}$ is related to the final film height, which is 25% smaller for the 3D case.

The kinetic anisotropic growth produces films of higher roughness, but the same initial state leads to a similar morphology under both growth modes (cp. Figs. 9.9 and 9.10). Therefore, the roughness is primarily related to the orientation of each grain's neighborhood in the initial state, which eventually leads to the formation of depressions.

3D simulations In the simulations of 3D film growth, periodic boundary conditions were applied in the two in-plane film dimensions (**y-z**-plane with 570×570 grid points), and isolation conditions on the top and bottom layers. 361 initial grains in the form of non-intersecting ellipsoidal caps aligned on $19^2 = 361$ square grid positions were used. As the seeding algorithm shown above allows only for a fixed number of orientations (in this case 337), 24 additional grains were initialized with random orientations, visible on irregular positions of the pole figure in Fig. 9.14. Three 3D scenarios were simulated and analyzed with the same distribution of orientations: coffin shaped crystals evolving by either surface energy or kinetic anisotropy (4 : 0.5 : 1 aspect ratio), and coffin shaped crystals with 4 : 4 : 1 aspect ratio evolving by kinetic anisotropy. The last growth morphology, reminiscent of a blade shape, was chosen to examine the influence of large in-plane shape anisotropy.

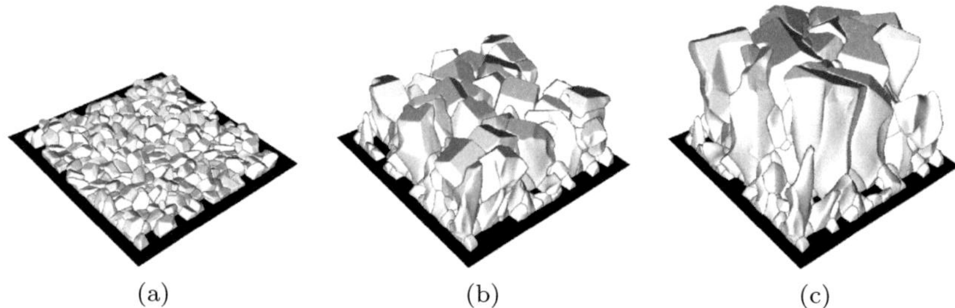

(a)   (b)   (c)

Figure 9.11: **(a)** - **(c)**: Three time steps in the evolution of the 3D zeolite thin film (kinetic anisotropy). Grains touching the lateral borders of the simulation box are omitted to reveal the internal grain boundary morphology.

Fig. 9.11 shows three time steps in the evolution of the 3D film (kinetic anisotropy) in an oblique view, where the iso-surfaces of the level set $\phi_\alpha = 0.5$ are rendered. Zeolite grains located at the lateral grid boundaries are left out to reveal an insight into the microstructure. Contrary to the 2D case, the internal grain boundaries are irregular and do not form flat planes during the growth process. Here, the interface dynamics is

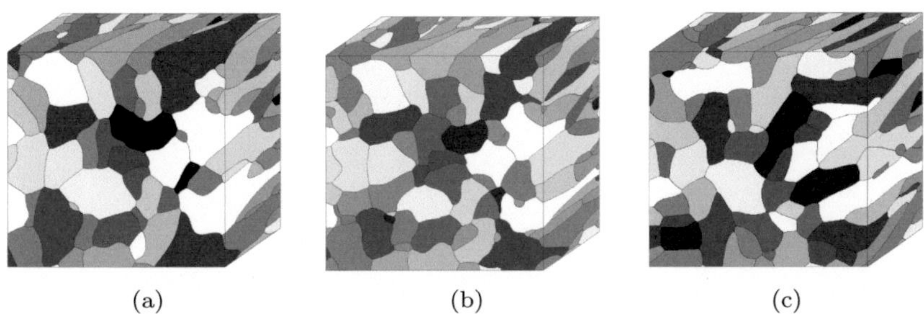

(a)   (b)   (c)

Figure 9.12: Sections of the film perpendicular to the growth direction, at half film width. Identical gray shades indicate same grains for **(a)** growth by kinetic anisotropy (61 grains), **(b)** growth under surface energy anisotropy (43 grains) and **(c)** crystals with $4:4:1$ aspect ratio kinetic Wulff shape (52 grains).

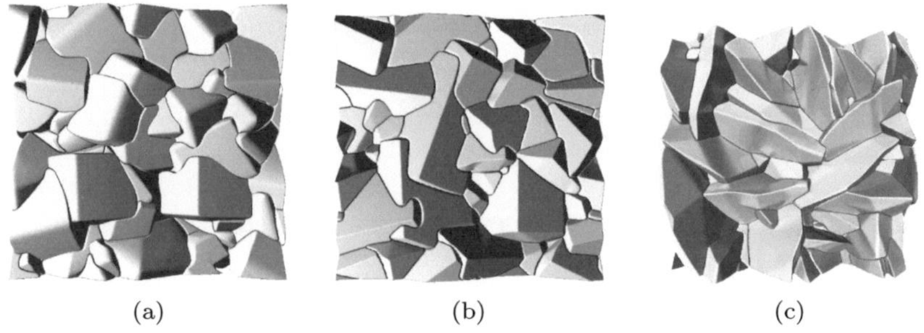

(a)   (b)   (c)

Figure 9.13: Top view on the 3D thin film briefly before it reached the upper simulation border (cube of side length $2.8\mu m$), with kinetic anisotropy **(a)**, surface energy isotropy **(b)** and blade like $4:4:1$ shape **(c)**.

modified by the effect of interface tension in the lateral (in-plane grain size) directions, not present in the 2D situation. In opposition to the 2D simulations, where all grain boundaries are along straight lines, in 3D the grain boundaries are formed by curved areas. A specific form of interfacial energy anisotropy between the solid grains, not specified for this system so far, could have a major effect, but is not included in the

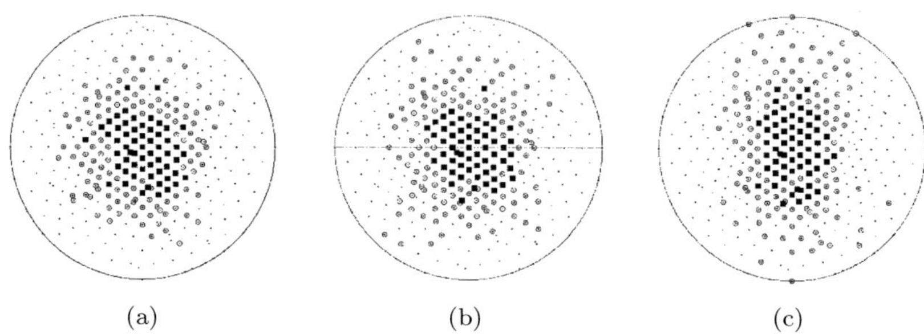

(a)　　　　　　　　　　(b)　　　　　　　　　　(c)

Figure 9.14: Stereographic projection of the evolution of grain orientations for 3D growth. Symbols are indicating the 361 orientations at simulation start (dots), after 50 % still growing (open circles) and 17 % of the grains still growing (solid squares), clearly showing the formation of a preferred orientation. Results for surface energy anisotropy (4 : 0.5 : 1 shape, **(a)**), kinetic anisotropy (4 : 0.5 : 1 shape, **(b)**) and blade shape **(c)**.

present simulations. Fig. 9.12 opposes isometric views of the fully grown film, planar sectioned at half film height, for all 3D simulations, showing the respective in-plane grain morphology. Comparing Fig. 9.12(a) and (b) reveals that, for the same crystal growth shape, kinetic anisotropy leads to a more regular in-plane grain morphology with smaller curvature of the grain boundaries and to a faster selection rate. As expected, the strong in-plane anisotropy of the blade-shaped crystals leads to a decisively different cross section in Fig. 9.12(c) with a high number of flat boundaries. All three morphologies exhibit elongated fibre-like grains in sections parallel to the substrate normal.

Similar selection processes as in 2D can be observed in competitive three-dimensional growth. In Fig. 9.13 a top view on the grown film at the final height is given for the different simulations. Again, orientations aligned close to the substrate normal are favored, and during time evolution, the other orientations become extinct. This can be visualized using a stereo projection of each growing orientation onto the $\mathbf{y} - \mathbf{z}$ equatorial plane in Fig. 9.14, where $\mathbf{x}$ is parallel to the film growth direction. The orientations still growing when 50% resp. 17% of the grains are left assemble around the center of the unit circle $\mathbb{S}^1$ in the case of the $1 : 0.5 : 4$ shape (Figs. 9.14a and 9.14b). The blade-like crystals represented in Fig. 9.14c seem to break this symmetry, as the mutual interaction during growth selects one azimutally preferred direction. In this case, the strong in-plane growth anisotropy obviously gives rise to a deviation from the fibre texture.

9.5 Discussion and outlook

This chapter presented the adaption of the general multi-phase field model of Sec. 6.2 to the problem of polycrystalline growth and the choice of the physical parameters. It was shown in detail, how the model can be used to study the texture evolution in thin films. The adaption of the model to the zeolite system comprises modeling the shape of the single crystals, which was studied using different solid-liquid interface anisotropies. This was done keeping in mind that most parts of the complex crystallization process in hydrothermal growth are simplified. The resulting growth process from seeds with uniformly distributed fixed orientations is dominated by selection of grains according to their orientation and a columnar morphology develops. Grains may survive in the growth competition, if they have the time to develop large enough facets at the sides of intruding neighbors. The preferential crystallographic orientation of the surviving grains (that are the grains that constitute the film surface) is parallel to the c-axis, the direction of fastest growth. The resulting microstructure depends only weakly on the strength of the driving force, but is strongly dependent on the actual initial orientations. To corroborate the results and to achieve a more quantitative validation, a comparison with experimental growth textures and microstructures has to be an essential part of the future work. Further, the differing solid-solid and solid-liquid interfacial free energies lead to a modification of the (liquid) dihedral angle with a possible impact on the triple junction motion and hence on the selection process. Also, anisotropy of the grain boundary energy may contribute to a net torque on the triple junction resulting from the Herring condition [147]. A systematic study of these effects and a quantification of their influence on competitive growth needed and shall be part of future investigations

The evolution of structure in thin films, namely size, shape and crystal orientation, is often discussed within the framework of structure zone models, which have been compiled from experimentally observed morphologies, mainly in physical vapor deposition processes [148]. In the simulations shown here, a low mobility of the grain boundaries between the growing crystals was combined with a comparatively high growth speed to represent the typical zeolite film growth process. This is a situation typical for zone 2 stage in the structure zone model, where the evolution is solely determined by kinetic factors, namely the different (constant) growth rates along different crystal directions. It was shown that this competitive growth mechanism, also known as 'van der Drift model' leads then, by pure geometric arguments, to a 'survival of the fastest', where in case of randomly oriented nuclei, grains oriented more or less parallel to the substrate normal will survive [149, 150]. There are various simulation studies in the literature for thin films based on constant facet growth rates [145, 146, 151, 152, 153, 154, 155], which corroborate this finding, but neglect the role of surface free energy.

The phase-field model applied in the present study, additionally takes into account interface thermodynamics and force balances at phase multi-junctions. In the simulations using a domain size within the range of experimentally grown zeolite films, at each stage of the growth process a maximum of the orientation distribution is found at the normal direction ($\theta = 0°$), i.e. an increasing development of the orientation of fastest growth, where the c-axis is perpendicular to the substrate. No other (temporary) maxima of the orientation distribution were observed in 2D growth, contrary to the simulation results by Bons and Bons using the same crystal shapes as in this study [137]. In the simulations, the overgrowth of less misoriented by stronger misoriented grains is a very rare event. The different force balances at liquid-solid triple junctions due to kinetic or surface energy anisotropy lead observably to different contact angles. Nevertheless, the overall selection dynamics and the resulting microstructure at later stages, namely the grain size and orientation distribution at the film surface remains substantially unchanged. In the phase-field simulations presented here oblique orientations are not subject to overgrowth only in the rare case of neighboring grains producing a fan-shaped arrangement. Existing experimental results reporting the evolution of oblique preferential orientations in zeolite growth are most probably due to a predominance of b (or $(0\ 1\ 0)$) facets in the seeding stage. In the colloidal seeding process on flat substrates, monolayers of seed crystals cover the substrate surface. Often, on the b facets growth twin crystals nucleate, with a relative rotation by $90°$ with respect to the c crystal axis of the parent crystal. This mechanism, which produces new orientations during film growth, is also an important factor in other poly-crystalline systems (e.g. poly-Si). This is part of ongoing studies. The key quantity which determines the final film shape is the initial orientation distribution, which is changed by the shape of the seeds and the morphology of the (in general rough) support. Concerning non-spherical seeds, the roughness of the support on a length scale equal to the dimension of the seeds and on a large length scale (non-flat support) are the decisive properties. The former changes the random orientation distribution, whereas the latter gives rise to a geometric screening, where regions protruding into the host phase are preferential. 2D simulations in a previous work corroborate the influence of the substrate [153].

Unexpectedly, the specific model of the growth anisotropy does not play a significant role on the resulting structure under the studied crystallization conditions. Grain competition follows clearly geometric arguments, orientation dependent force equilibria at triple junctions do not make a significant change in the studied growth regime. The film develops an essentially closed surface with eventual singular clefts, in the general case the roughness is determined by the mean in-plane grain diameter and the growth facet inclination.

Furthermore, all solid-solid boundaries exhibited no remaining liquid phase using the simulation parameters defined in Sec. 9.4. To examine a possible effect on the film morphology, different driving forces for crystallisation were tested for several simulations in 2D, ranging from 10% to 200% of the reference value. In case of surface energy anisotropy and low driving force for crystallization, a thin layer of liquid wets some of the grain boundaries, as the solid-liquid $(1\ 0\ 0)$ crystal facet has a much lower energy than the grain boundaries. This effect has substantially no influence on the speed of orientation selection or on the resulting film shape. Nevertheless non-closed films and mesoscale porosity could remain as a consequence, as crystal growth would slowly proceed in (partly) isolated liquid domains.

In this study, transport limitation of the silicate material within the liquid phase was assumed to be of minor importance. Zeolite crystallization can be explained as a process involving the attachment and reorientation of nanoscale silicate building blocks as the rate determining step (see e.g. the review of Cundy and Cox [156] and the references there in). The general phase-field model allows for the easy incorporation of multi-species diffusion [157] and growth kinetics, which may depend on local concentration. It is possible that under special conditions preferential orientations not perpendicular to the substrate evolve, as it can be found in the directional solidification of alloys. This issue will be explored in future work, after more precise data on the interface properties of zeolite crystals and free energies of the involved phases has been collected.

10 Phase-field modeling of the magnetic shape memory effect

The phase-field model introduced in Chap. 6, Sec. 6.2 is now applied to simulate effects related to the magnetic shape memory effect in the Heusler alloy Ni_2MnGa. This chapter is based in its most parts on three articles that are published or accepted for publication in an international journal: First, a contribution published in the proceedings to the Fifth International Conference on Multiscale Material Modeling, held in October 2010 in Freibug, Germany [4]. Second, an article that appeared in the Archives of Mechanics in 2011 [5]. And third, an article the authors were invited to publish after an oral contribution at the Joint European Magnetic Symposia, held in September 2012 in Parma, Italy [8]. This chapter follows these articles in structure and text.

As Chap. 5 stated, the modeling of MSMAs and their properties is a very challenging task. Models for the MSME, the elastically and magnetically induced rearrangement of martensite twin boundaries and magnetostrictive processes have been published, among others, by deSimone and James [11], Kiefer and Lagoudas [158], Miehe et al. [159] and Conti et al. [160]. There exist several approaches to model the MSME that are based on the phase-field method, published e.g. by Jin [58], Zhang and Chen [88], Landis [97], Li et al. [161] or Mennerich et al. [5]. These models describe the effect on the mesoscale, but significantly differ in the choice of the order parameters and employed potentials. The common aim is the computation of magnetization vs. magnetic field or stress vs. strain behavior to gain an understanding of the fundamental processes leading to the MSME and related processes to render possible the prediction of the behavior of materials providing the effect. Simulation results and computed curves can be compared to experimental results to analyze hysteresis behavior, as done by Arndt et al. [42] or Krevet and Kohl [162] using non-phase-field approaches. Phase-field models for the field of domain evolution in ferro-electrics are developed by Su and Landis [163] or Schrade [164].

10.1 Simulation setup and parameters

Several simulations were carried out to study the microstructure rearrangement in near tetragonal Ni_2MnGa in the modulated 5M state. Both, the effect of applying external elastic forces and magnetic fields are analyzed. Main attention in the simulations is drawn to understand the evolution dynamics and transition pathways of the martensite rearrangement, and to compare reached steady state results to those predicted by theory and experiment. The following assumptions are made: First, the operation temperature T_{op} is below the Curie temperature T_{Curie} and the martensitic start temperature T_{ms} (i.e. $T_{op} < T_{Curie}, T_{ms}$). Second, any externally applied magnetic field is constant over (sufficiently long periods of) time.[1] Third, the material under consideration has to be ferromagnetic hard and homogeneous (in the sense that the concentration is the same everywhere in the material).

The simulation domains are rectangular boxes with a regular grid in 3D. The evolution of all three components of the spontaneous magnetization $\mathbf{m}$ and displacement field $\mathbf{u}$ were calculated (cp. Chaps. 3 and 4). To save computation time, mostly quasi 1D and 2D settings were used, in which the magnetization and the elastic displacement field are still free to evolve in all three spatial dimensions. For the field $\mathbf{u}$, either fixed displacements or surface traction forces can be applied at the boundaries, for the magnetization either the special Neumann condition $\frac{\partial \mathbf{m}}{\partial \mathbf{n}} = 0$ or periodic boundaries to represent an RVE can be assumed.

The MSME problem includes the interdependent evolution of twin domains and magnetic domains with dimensions and interfaces spanning

---

[1] This is necessary for the minimization procedure for the spontaneous magnetization $\mathbf{m}$ that is described by Eq. (4.8). Formally, the *Liapounov structure* of the system has to be maintained (see [56]). That way, no eddy currents are induced.

different length scales. This has to be taken into account carefully when a suitable parameter set is chosen. In the following, the magnetic properties of 5M tetragonal Ni_2MnGa are given in SI-units. The saturation magnetization is chosen as $M_S = 6.015 \cdot 10^5 \frac{A}{m}$, and the magnetocrystalline anisotropy constant as $K_{\text{aniso}} = 2.45 \cdot 10^5 \frac{J}{m^3}$ (taken from [165]). The magnetic exchange constant is chosen as $A_{\text{exch}} = 2 \cdot 10^{-11} \frac{J}{m}$ (see [88]). The width $\delta = \sqrt{\frac{A_{\text{exch}}}{K_{\text{aniso}}}} = 9 \cdot 10^{-9} \, m$ of Bloch walls is a typical transition scale between the magnetic domains (cp. [44]). This has to be resolved on the numerical grid, leading to the choice of the physical grid distance to be $\Delta x = 2 \, nm$. This results in a 2D simulation domain of $1 \, \mu m \times 1 \, \mu m$ at 500 grid points resolution as a typical physical size of the simulated material volume. The parameters entering the micromagnetic evolution Eq. (4.8) are chosen according to [96]: A gyromagnetic ratio of $\gamma = 2.21 \cdot 10^5 \frac{m}{As}$ and a damping factor of $\alpha_G = 0.5$ are used. To treat the elastic problem in Ni_2MnGa the mass density of $\rho = 8.02 \frac{g}{cm^3}$ is used (as in [165]). The tetragonal elastic stiffness tensor of the martensite variants is approximated by an averaged cubic tensor with values from [90] (cp. Tab. 10.2), so that homogeneous cubic symmetry is assumed in the solution of the elastic Eqs. (7.2) or (7.6). The crystallographic data for the transformation strains of the tetragonal martensitic variants are taken from [166] and the transformation matrices are of the form given in Eq. (5.3). Only the diagonal components have non-zero values of $\alpha = 0.019$ and $\beta = -0.041$, so that the c-axis of variant V_1, represented by U_1, points in the **x**-direction, the c-axis of V_2 in the **y**-direction. For the simulations, equations are non-dimensionalized. The dimensionless quantities are indicated by a tilde and the scaling factors by the subscript zero. Therefore, spatial coordinates are expressed by $r = \tilde{r} \, d_0$ using the length scale $d_0 = 2 \, nm$. A time-scale is fixed as $t = \tilde{t} \frac{1}{\gamma M_s} = \tilde{t} \, t_0$ with $t_0 = 7.52 \cdot 10^{-12} s$. Together with a typical magnetostatic energy scale $f_0 = \mu_0 M_s^2 = 4.55 \cdot 10^5 \frac{J}{m^3}$, all bulk energy terms in the functional Eq. (6.5) can be written dimensionless. From the relation between magnetic field and energy (see Eq. (4.9)), the magnetic field scaling factor is then fixed as $H_{\text{eff},0} = M_s$. In Tabs. 10.1 and 10.2, the physical parameters are shown, together with their dimensionless values that were used throughout all simulations concerning the magnetic shape memory effect. Additionally, for the interface tension of the twin boundary in the phase-field equation Eq. (6.10), a value

Table 10.1: Magnetic parameters for Ni_2MnGa, which were used in the simulation of twin boundary motion, including the non-dimensional values using the length scale $d_0 = 2nm$, time scale $t_0 = \frac{1}{\gamma M_s} = 7.52 \cdot 10^{-12} s$ and the energy scale $f_0 = \mu_0 M_s^2 = 4.55 \cdot 10^5 \frac{J}{m^3}$, with $\mu_0 = 4\pi \cdot 10^{-7} \frac{N}{A^2}$ being the permeability of vacuum.

quantity	M_S	K_{aniso}	A_{exch}	γ
SI units	$\frac{A}{m}$	$\frac{J}{m^3}$	$\frac{J}{m}$	$\frac{m}{As}$
value	$6.02 \cdot 10^5$	$2.45 \cdot 10^5$	$2 \cdot 10^{-11}$	$2.21 \cdot 10^5$
dim-less	1.0	0.539	1.76	1.0

Table 10.2: Elastic and twin interface parameters for Ni_2MnGa, including dimensionless values.

quantity	ρ	c_{11}	c_{12}	c_{44}	γ_{twb}
SI units	$\frac{g}{cm^3}$		$\frac{J}{m^3}$		$\frac{J}{m^2}$
value	8.02	$1.60 \cdot 10^{11}$	$1.52 \cdot 10^{11}$	$0.43 \cdot 10^{11}$	0.1
dim-less	1694	$3.519 \cdot 10^5$	$3.343 \cdot 10^5$	$0.935 \cdot 10^5$	47.5

of $\gamma_{\alpha\beta} = \gamma_{twb} = 0.1\frac{J}{m^2}$ is assumed for each interface α/β (cp. Eqs. 6.7 and (6.9)). This value is more than an order of magnitude smaller compared to typical grain boundary interfacial tensions. This value is not well defined in the literature, but can in principle be calculated from the atomistic variant structure by ab initio methods. The diffuse interface width for the phase fields was taken as $\xi = 3\,d_0$, what is slightly smaller than the magnetic transition width and results in a resolution of about 8 grid points on the numerical grid. In the simulations, the kinetic coefficient in Eq. (6.10) was set to $\tilde{\tau} = 1$, so that $\tau = \tau_0 = \frac{f_0\,t_0}{d_0}$. Here we expect that the interface velocity is not significantly modified by the order parameter evolution, but is dominated by the kinetics of strain propagation and magnetic evolution. For the case of elasticity the time evolution is related to material density, elastic coeffcients and the damping coefficient in the wave equation (7.2), for which a value of $\tilde{\kappa} = 1000$ was chosen.

Ni$_2$MnGa magnetic shape memory alloy single crystals are typically operated under compressive stress along one of the variants c-axes and under an external magnetic field in the perpendicular direction. Because the martensitic variant with the short c-axis along the direction of compression minimizes the elastic energy, and a second variant with this direction oriented along the external field minimizes the magnetic energy, a two-variant state is favored when the system minimizes the free energy. The simulations that were carried out in this chapter approach this situation step-by-step. The following simulation scenarios consist of two variants V_1 and V_2, where the index is representative for the Bain strain given in Eqs. (5.3). To visualize the direction of the spontaneous magnetization and the formation of magnetic domains, either arrows are used or a color coding scheme that shows moments that are aligned with a variants easy axis in lighter, and moments that are aligned anti-parallel in darker shades.

10.2 Hysteresis in Ni$_2$MnGa

The first simulations shown in this chapter are pure micromagnetic simulations that were set-up to analyze the hysteresis behavior in a Ni$_2$MnGa single crystal consisting of only one single martensitic variant, and to compare the outcome with the results published by Tickle and James in 1999 [165]. The parameters were taken as introduced in the last section. The simulation box had the dimension $100 \times 1 \times 100$ and represents an RVE taken out of a surrounding spherical specimen, assuming a physical length scale of $\Delta x = 10$nm. The RVE was initially non-magnetized. Two magnetization cycles were performed, one in the direction of the variants easy axis (that coincides with the tetragonal c-axis), and one in the direction orthogonal to the easy axis (the so called *hard axis*). In each cycle, the external magnetic field was successively increased in small steps, and the magnetization was in a steady state between each two steps. The resulting magnetization vs. external field curves are shown in Fig. 10.1. As can be seen, the curves compare quite well to the measurements of Tickle and James [165], saturation is reached at similar external field strengths. When saturation is reached, the external field is reversed again, and in the case of magnetizing the sample in hard axis direction this leads to

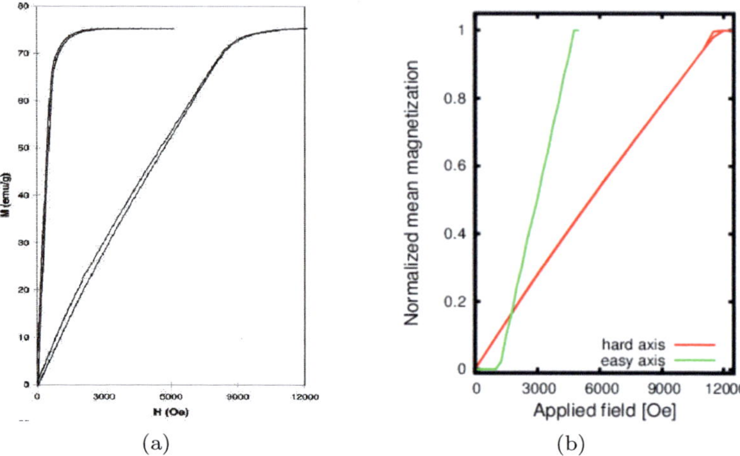

(a)   (b)

Figure 10.1:   **(a)** Experimental measurements of hysteresis in a Ni_2MnGa single variant crystal published by Tickle and James [165] and **(b)** micromagnetic simulations performed in an RVE of a Ni_2MnGa single variant. In general, the results compare well, but no hysteresis is achieved in the simulation of the magnetization in easy axis direction.

hysteresis behavior. The deviation of magnetic moments from the easy axis exceeds the Zeeman energy, and the moments turn out of the hard axis direction again. In opposition, however, no hysteresis can be achieved in the simulations when the sample is magnetized in the direction of the easy axis, as no nucleation mechanism for the martensitic variants or magnetic domains is included in the numerical calculations. The strength of the magnetocrystalline anisotropy energy in the Ni_2MnGa specimen is, in combination with the magnetic exchange energy, too strong to let the demagnetization field take effect and demagnetize the sample.

10.3 Accomodation of external strain

The accomodation of externally imposed strain (neglecting micromagnetic forces) was studied with the phase-field model using the parameters introduced in Sec. 10.1. A periodic RVE was assumed, and a domain of 64×64 grid points initialized with random values for the phase fields of two variants V_1 and V_2, giving random volume fractions locally (see Fig. 10.2a). Different homogeneous strains, reflecting the volume fraction of V_1 as $v_1 = 0.5$, $v_2 = 0.6$, $v_3 = 0.7$ and $v_4 = 0.8$ were applied reflected by the homogeneous strains

$$\bar{\epsilon}_1 = \begin{pmatrix} -0.011 & 0 & 0 \\ 0 & 0.019 & 0 \\ 0 & 0 & -0.011 \end{pmatrix}, \bar{\epsilon}_2 = \begin{pmatrix} -0.017 & 0 & 0 \\ 0 & 0.019 & 0 \\ 0 & 0 & -0.005 \end{pmatrix},$$

$$\bar{\epsilon}_3 = \begin{pmatrix} -0.023 & 0 & 0 \\ 0 & 0.019 & 0 \\ 0 & 0 & 0.001 \end{pmatrix}, \bar{\epsilon}_4 = \begin{pmatrix} -0.035 & 0 & 0 \\ 0 & 0.019 & 0 \\ 0 & 0 & 0.013 \end{pmatrix},$$

The boundary conditions Eq. (7.8) were applied and a mechanical equilibrium (see Eq. (7.6)) was assumed. The evolving volume fraction of V_1 was recorded (cp. Fig. 10.2). Initially, from the random structure a non-branched lamellar arrangement of V_1 and V_2 quickly develops, which finally takes a volume fraction of 50%. Twin boundaries along the expected $\langle 110 \rangle$ crystal directions appear, similar to the results depicted in Fig. 10.3. The successive application of $\bar{\epsilon}_i = v_1 \epsilon_0{}^1 + (1 - v_i)\epsilon_0{}^2$ results in volume fractions numerically very close to the expected values.

10.4 Periodic boundary conditions and RVEs

The usage of periodic boundary conditions to mimic infinite extended periodic structures as RVEs opposes restrictions on the developing variant and magnetic domain structures. This is briefly analyzed here. Both, the magnetization and the variant structure are enforced to be periodic. Furthermore, the magnetic domains are additionally bound to the variants easy axes when no external magnetic field is present. The imposed periodicity may affect the emerging structure. Fig. 10.3 shows an example of calculations in a 60×60 domain, where an initially random distributed two-variant structure (analogous to the one shown in Fig. 10.2a) evolves under the boundary condition $\sigma^{\mathrm{appl}} = 0$ in Eq. (7.10) by using the equilibrium Eq. (7.6). The simulation starts from this state with an additionally random distributed magnetic structure. The final equilibrium state consists

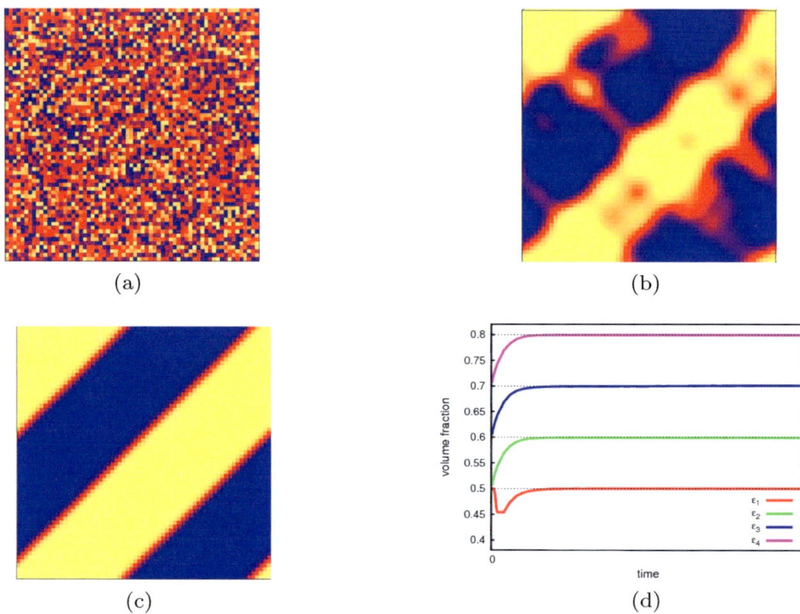

Figure 10.2: **(a)** Randomly distributed two variant structure, **(b)** decomposition of the initial structure into a lamella structure and **(c)** the resulting laminate after application of homogeneous strain ϵ_1. The volume fraction of variant V_1 (shown in yellow) is 0.5. **(d)** Development of volume fraction of variant V_1 starting from a random equi-distribution of two variants in the domain and successive application of homogeneous strains ϵ_1 to ϵ_4 that reflect volume fractions of V_1 to be 50%, 60%, 70% and 80%, respectively. The dotted horizontal lines indicate the expected volume fraction.

of a single variant and two domains with vanishing total magnetization, a consequence of the action of the demagnetization field and the contributions from $N\bar{\mathbf{m}}$ in Eq. (8.10) (where $\bar{\mathbf{m}}$ denotes the average magnetization in the simulation area). The reason for the structural change lies in the development of an unpreferential domain structure in early stages of the evolution process, as the magnetization is forced to respect the periodicity of the system. The topological constraint enforces branching of magnetic domains (see lower and upper left part of Fig. 10.3a). This leads to head-to-head and tail-to-tail boundaries at the twin interfaces with high exchange energy that cause the variant structure to vanish. One has to bear in mind here that the use of a stress boundary condition that is realized by setting $\boldsymbol{\sigma}^{\mathrm{appl}} = 0$ in Eq. (7.10) implies the existence of a small sample with the size of the simulation box. The results of a second sim-

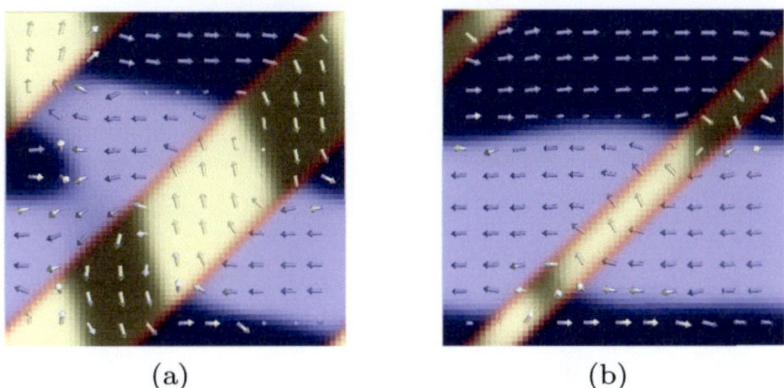

(a)   (b)

Figure 10.3: Interaction between magnetic domains and variant structure: **(a)** Formation of a lamella variant structure from an initially random phase and magnetic structure. By chance, unfavorable branched magnetic domains evolve due to the periodic boundary constraint. Energetically unfavored head-to-head and tail-to-tail domain boundaries can be seen where lighter and darker shaded parts meet. **(b)** Unfavorable magnetization states exert a force on the twin boundaries, such that one variant vanishes.

ulation that started with zero applied magnetic field $\mathbf{H}_{\mathrm{ext}}$ and a variant lamella structure of 50% of variant V_1 and V_2 are shown in Fig. 10.4. A periodic structure with two magnetic domains develops, where only 90° and 180° domain walls form. This is the pattern commonly observed in experimental work and often used for an analytical description of the magnetic field induces strain. When the magnetic structure has become stable, an external magnetic field of about 250mT in the easy axis direction of variant V_1 (yellow) is applied, favoring this variant. It can be seen that the magnetic domain structure dissolves, and due to the increase of Zeeman energy of V_2 (blue) the twin boundaries move. As wrinkles in the interface would strongly increase the elastic misfit energy, the interfaces between the variants stay straight during their motion. It is noteworthy that the periodicity in this simulation imposes that each two twin boundaries are connected and move as a whole.

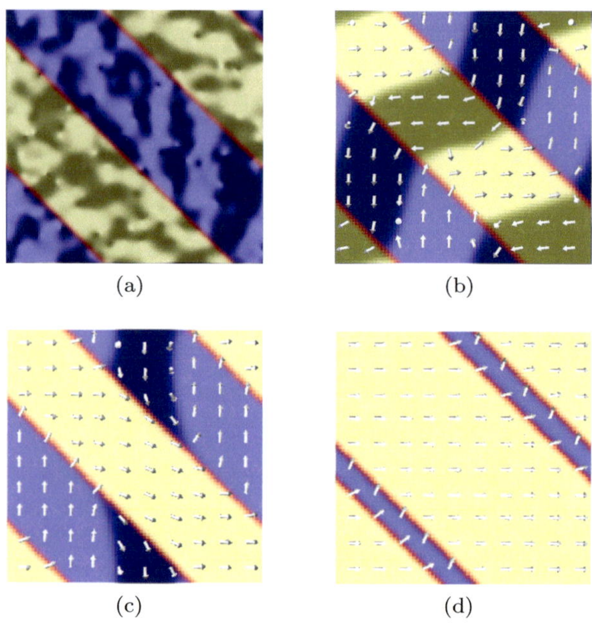

Figure 10.4: Formation of magnetic domains in Ni_2MnGa: **(a)** Initial lamella structure and formation of magnetic domains from an initially random magnetization state. **(b)** Equilibrium magnetization state that respects the periodicity of the RVE. **(c)** Application of an external magnetic field (pointing to the right) that causes the domains to dissolve quickly within the growing variant. In the shrinking variant, domain walls can be observed for longer times. **(d)** Motion of twin boundaries to minimize the Zeeman energy of the unfavored variant.

10.5 Magnetic field induced strain

Simulations were carried out to analyze the magnetic field induced strain effect under typical operation conditions as stated above. That means, the behavior of an MSM material is simulated under the application of concurrently acting external applied stresses and magnetic fields. A periodic domain of 80×80 grid points was used. All spatial components of the elastic displacement field and the spontaneous magnetization are taken into account. The domain contains a two variant laminate showing 50% of each variant V_1 and V_2. It is nearly non-magnetized due to a periodic magnetic domain structure consisting of $90°$ and $180°$ domain walls. The sample is assumed to be included in a spherical uniformly magnetized specimen, so that the shape factor can be assumed to be $N_{\text{sphere}} = \frac{1}{3}\mathbf{I}$ (see Sec. 8.2.2). This initial setting is analogous to the one shown in Fig. 10.4b. Starting from this state, an external magnetic field $\mathbf{H}_{\text{ext}}$ in the direction of the c-axis of variant V_1 was applied, ranging from -550 mT to 550 mT in constant steps. The average strain and magnetic moment in the direction of the field were measured. The numerical experiment was carried out once without an applied load, and second with a compressive load of $\sigma^{\text{appl}} = 0.5$ MPa (which is below the twinning stress σ_{tw}, cp. the beginning of Sec. 10.6) orthogonal to the direction of the magnetic field. The resulting stress and magnetization curves are shown in Fig. 10.5. Each curve consists of a total of 110 successive single simulations, each of which has been conducted long enough to reach a steady state in the magnetization. A continuous rate is not feasible due to the small numerical step width necessary to solve Eq. (4.8). In the actual implementation of the model no mechanisms that allow for the nucleation of martensitic variants is included. Hence, the maximum external field $\mathbf{H}_{\text{ext}}$ was chosen not to be strong enough to completely saturate the sample to a single magnetic domain, and the field was reversed before the specimen was in a single martensitic variant state. As the simulations start in a non-magnetized state, the initial $180°$ domain walls dissolve in the first stages of the magnetization process (comparable to the simulation shown in Fig. 10.4c), and no motion of twin boundaries occur. When the the external magnetic field exceeds a value of about 300 mT under no load (and about 400 mT in the compression experiment), V_2 transforms to V_1. Closely before saturation is reached, the field is reversed at 550 mT. The sample stays in an almost constantly strained state until the external field drops below the value of about 210 mT. In this regime the external field dominates the demagnetization field. Then, driven by minimizing the demagnetization energy, the respective field $\mathbf{H}_{\text{demag}}$ turns the magnetization so that V_2 is energetically favored. The sample shows a remanence magnetization and non-zero strain at zero magnetic field, what is different from experimental measurements (see e.g. [57]). This effect can be attributed to the absence of sample boundaries of the periodic RVE, typically a source for the nucleation of new 180° domains (see [44]) and to the assumption of a uniformly magnetized sample for Eq. (8.10). The demagnetization process continues until the field, now pointing in the opposite direction, is strong enough to demagnetize the sample. This is the coercive field at about -200 mT, where the the magnetization switches its sign. Now, again variant V_1 is favored and grows, leading to an increase of strain.

10.6 Dynamic loading behavior

When MSMAs are used as actuators the value of the twinning stress $\boldsymbol{\sigma}_{\text{tw}}$ is of special interest. This is the threshold to be overcome to induce the motion of twin boundaries. In the literature, different experimental values ranging from $0.7 - 4$ MPa [57] up to 20 MPa [167] are reported for Ni_2MnGa, and an influence of the loading rate can be expected.

First analysis was taken towards an understanding of the influence the interface relaxation parameter τ in the phase-field equation Eq. (6.10) and the damping parameter κ in the damped wave equation Eq. (7.2) have on the simulations concerning the MSME. To study the model under dynamic mechanical conditions, simulation series were carried out to analyze the stress vs. strain behavior at finite applied stress rates under applications of uniaxial tensile stress. For this purpose, the magnetic energy terms were switched off. A quasi one-dimensional box of 100 grid points with boundary points on all sides was used, i.e. the contributions in all spatial directions were taken into account. As in this case the variant boundary need to be perpendicular to the long direction of the box, as imposed by periodicity, the underlying crystal was rotated by 45°. The elastic property tensor, the variant's easy axes and the applied stress were transformed accordingly. The applied stress was increased linearly in time up to a final level of 4.5 MPa, giving a physical loading time of $6.8 \cdot 10^{-7}$ s, resulting in a shock loading rate of $6.7 \cdot 10^6 \frac{\text{MPa}}{s}$. The variant volume fraction was chosen to be 20% in all simulations, and this state was declared as the reference unstrained state. Hence, the maximal achievable amount of recoverable strain of 6% cannot be reached in this simulations. The strain evolution was calculated using the dynamic equation Eq. (7.2) with different values for the damping coefficient κ, which can be reformulated as the quotient of the mass density and the time scale t_{drag}, by which the motion is slowed down. The effect of the term $\kappa\dot{\mathbf{u}}$ in the dynamic equation (7.2) is to introduce a dissipation mechanism in conjunction with a drag force acting everywhere in the bulk. As a reference, the parameters $\tau = 16$ and $\kappa = 1000$ were arbitrarily taken. The resulting stress vs. strain curves for damping values of $\kappa = 1000, 1500, 2000, 3000$ and 4000 are given in Fig. 10.6. The material deforms mainly linearly elastic below a stress level of $\sigma_{\text{tw}} = 3.25$ MPa, independent of the damping coefficient. This value is quite close to the theoretical limit value of $\frac{K_{\text{aniso}}}{\epsilon_0} = 4.0$ MPa for the parameter values used in this study (cp. [57]). Above σ_{tw} a plateau with small slope indicates the accommodation of strain by rearranging the twin boundaries, known as the super-elastic effect. After this stage at a strain of 4.8% the material is completely transformed to a single variant, and the material again behaves linearly elastic. The threshold σ_{tw} where twin boundary motion is induced, does not change significantly. The apparent elastic modulus of the initial variant mixture is smaller by a factor of two compared to the modulus of the final single variant state, as there is already a small and constant transformation rate from V_1 to V_2 before reaching the critical stress σ_{tw}. The coefficient τ in the phase-field Eq. (6.10) incorporates the interface kinetics and relates to an interfacial drag force term of size τv_n in the sharp interface limit ($\xi \to 0$), where v_n is the interface normal velocity. When the behavior of the phase-field model in the limit of a thin finite interface width is studied analytically, it can be used to quantitatively establish a physical relation between the driving force of the transition and v_n [78]. The driving force in the problem at hand is the difference in elasto-magnetic energies across the variant boundary. For this model, no such analysis is available to date. Hence, a linear relation between driving force and velocity as expressed in Eq. (6.5) is postulated. This requires the evolution of the phase fields (representing the order parameters for the variant eigenstrains) not to slow down the transition. A quantitative scaling of the relaxation parameter τ to integrate kinetic laws of the twin boundary motion as have been recently published by Faran and Shilo [168], has to be a part of future works. Fig. 10.6b shows the result of a series of simulations, where τ was increased by successively doubling its value starting at $\tau = 16$, while $\kappa = 1000$ was fixed.

Smaller values of τ represent higher relaxation rates (cp. Eq. (6.10)). Values above $\tau = 128$ have not been studied, as they led to elastic oscillations in the vicinity of the interface, giving rise to an oscillatory growth velocity that resulted in unstable numerical simulations. Again, the twinning stress is found at 3.25 MPa independent of interface kinetics, and the slope of the stress plateau increases with increasing interface drag τ, but shows a very weak dependency.

10.7 Three variant state in 3D

The simple laminar configuration consisting of two martensitic variants V_1 and V_2 in 2D as e.g. shown in Fig. 10.2c was extended in the **z**-direction for a cubic 3D domain resolved by a grid size of $320 \times 320 \times 320$ grid points. The third martensitic variant V_3, having its short tetragonal c-axis along the **z** direction and thus orthogonal to those of the other two variants, was placed atop at about $\frac{2}{3}$ of the height of the simulation box. Compressive stress of 1.13 MPa (via surface traction boundaries) was applied along the **x**- and **y**-directions, which are the directions of the long crystal axes of variant V_3.

The values for the eigenstrain tensors were altered to $\alpha = 0.01$ and $\beta = -0.02$ according to [58], and the surface tension to $\gamma_{twb} = 0.018\frac{J}{m^3}$. The initial magnetization was set parallel to the $\langle 111 \rangle$ diagonal. Periodic boundary conditions for the magnetization were used, for the phase field parameters periodic boundaries in the **x-y**-plane and special Neumann boundary conditions in the out-of plane dimension were applied. Because of the computational complexity of this simulation, the demagnetization energy was neglected by explicitly setting $\mathbf{H}_{\mathrm{demag}} = \mathbf{0}$ in Eq. (4.10), underlining that this simulation is a test to show the general aplicability of the model presented here. In Figures 10.7a and 10.7b the isosurfaces of the phase fields of V_1 and V_2 at an intermediate value of the order parameter, $\phi_\alpha = 0.5$ and $\phi_\beta = 0.5$, are shown for an early and a later timestep. The magnetic domain structure is not shown. During the evolution, an intricate interface between the V_1-V_2 laminate and the third variant forms, consisting of zig-zag shape arrangements of (110) facets as shown in Fig. 10.7a. As expected, V_3 dissolves and completely vanishes.

In the later stage, twinned platelets grow into V_3, starting from the edges of the roof-like V_1-V_2 twin laminate surface.

10.8 Discussion and outlook

A general phase-field model was adapted to model the process of twin boundary motion in the martensitic state of the shape memory alloy Ni_2MnGa in the near tetragonal 5M state, based on the interpolation of elastic and micromagnetic free energies. Periodic boundary conditions to mimic RVEs are used to describe the transformation process of a martensitic laminate by external mechanical and magneto-mechanical load. Strain accomodation and stress-strain behavior with present or absent external magnetic fields show good agreement with experimental results.

Two numerical issues limit the length and time scale of the simulations severely: First, the stable solution of the magnetization dynamics involves very small time updates in the order of 10^{-14} s, so that periods of about microseconds are computationally accessible. Here, adaptive time stepping or similar techniques and the solution of the dynamic Eq. (4.8) in the overdamped limit $\alpha_G \to \infty$ could be a cure. Second, the numerical resolution of the magnetic domain boundaries with widths in the ten nanometer range limits the physical domain sizes. This drawback could be overcome by the use of adaptive meshing techniques.

To enable the examination of complete stress-strain hysteresis cycles in the simulations nucleation mechanism for the martensitic variants are to be included in the model description. This can either be done be including a stochastic noise term of definite amplitude and distribution into the phase-field functional $\mathcal{F}$ (see Eq. (6.5)), or by explicitly inserting martensitic nuclei in the calculation domain.

In the phase-field model used here, the interfacial tension of the variant boundary enters the surface energy terms $a(\phi, \nabla\phi)$ and $w(\phi)$ in Eq. (6.5) as a proportionality constant (cp. the definitions in Sec. 6.2). Due to the interpolation of free energies over a finite interface width an additional contribution to the interface free energy may arise. The conducted simulations show that the phase-field profile ϕ_{V_1} has a width that is smaller than expected, which points to the existence of an undesired interface excess, as the difference in the free energy of adjacent martensitic variants enters the equilibrium width of the interface that separates these variants. A detailed analysis of its influence applying similar techniques as presented by Choudhury and Nestler [78] is a necessary part in the future the work in this field. This includes detailed simulation studies of the influence that changes in the value of the surface tensions parameter $\gamma_t wb$ and the parameter ξ that effects the width of the diffusive interface have on the energetics of the system. First preliminary analysis of simulation data leads to the suggestion that the critical stress level $\boldsymbol{\sigma}_{\mathrm{tw}}$ is affected by a change in the parameter $\gamma_t wb$, but this has so far not been subject of more detailed analysis.

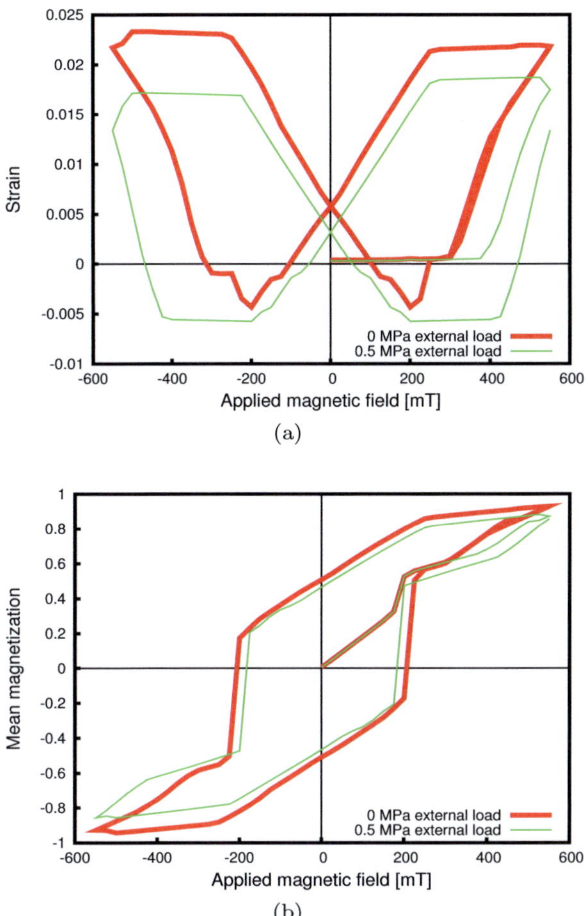

Figure 10.5: Magnetic field induced strain, starting from a two-phase lamination with initially 50% volume fraction of two phases under different loads of 0 MPa and 0.5 MPa. The initial state is defined as the unstrained reference state: **(a)** Strain vs. an external applied magnetic field. The curve shows the typical butterfly shape. As the external magnetic field acts against the applied external load, the transition starts later and the achievable strains are lowered when an external stress is applied. **(b)** Mean magnetic moment vs. external applied magnetic field measurement results for the same to simulations. The curve shows a remanence magnetization of the sample, and a coercive field of about -200 mT.

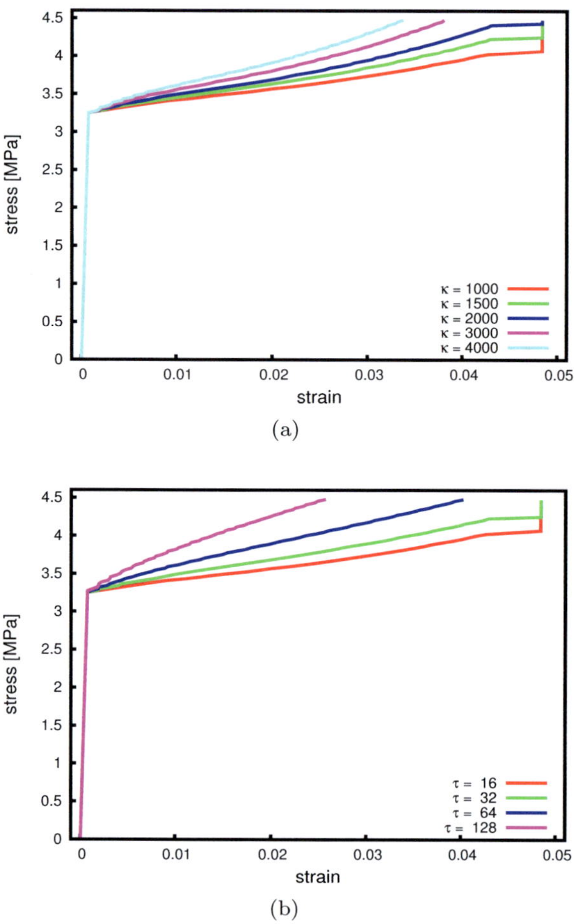

Figure 10.6: Simulation data for the stress vs. strain relationship in a quasi 1D sample: **(a)** Variation of damping coefficient κ at constant interface relaxation coefficient $\tau = 16$ and **(b)** variation of interface relaxation coefficient τ at constant damping $\kappa = 1000$.

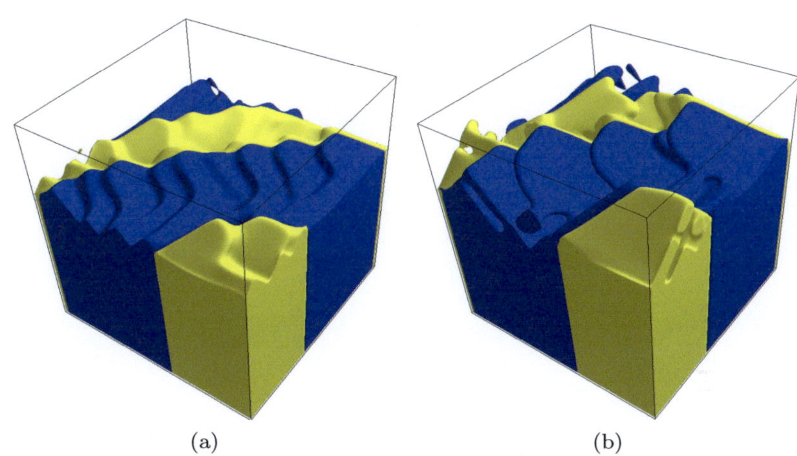

(a)   (b)

Figure 10.7: Simulation of three different martensitic variants in 3D. Only two variants are shown, the third, not shown variant is atop the other two. **(a)** Plot of the isosurfaces $\phi_{V_1} = 0.5$ and $\phi_{V_2} = 0.5$ in an early stage of the evolution process, where facets start to form. **(b)** Final stage of the evolution before the top boundary of the simulation box is reached. The developing (110) facets can be observed.

11 Outlook

This concluding chapter addresses some general topics related to the model proposed in this work, and discusses possible future modifications and applications. The most important aspects directly linked to the simulation studies presented in the last two chapters have been discussed there in the corresponding outlook sections.

The model developed in the context of this work has proven well in different scenarios, and opens up for further applications. Special attention in this work was paid to the development of a model description that couples a phase-field approach with micromagnetics and mechanical elasticity. The main focus was drawn on developing computation methods to make the micromagnetic problems feasible, on finding sound magnetic and elastic boundary conditions and on solving the elastic equations (the dynamic wave equation Eq. (7.4) and Eq. (7.6) for the mechanical equilibrium) in a general context, allowing, in principle, for arbitrarily oriented phases with differing elastic properties. Nevertheless, when the elastic dynamic wave equation is solved that is implemented at the moment, it is hard to relate the damping mechanism to energy dissipation properties and match it with physical conditions (cp. the parameter κ in Eq. (7.2)). Finel et al. include the kinetic energy density and a *Rayleigh dissipation density* into their modeling approaches (see [169]). Applying theses ideas might improve the phase-field simulation results that are achievable with the model presented in this work. Further, the restriction to linear elasticity when modeling the MSME, although often used, is considered a severe limitation sometimes in the literature (cp. e.g. [169]): The disregard of large deformations is indicated as a source of non-physical behavior, because the giant strains attributed to pseudoplastic behavior are a characteristic of MSMAs. It has to be investigated if, in the context of ferromagnetic shape memory alloys, the geometric linearization of elasticity is justified, or if a model formulation based on non-linear elasticity leads to more appropriate results. The following sections briefly sketch some possibilities of future applications for the developed models.

11.1 Magnetic domains in magnetic shape memory alloys

An interesting application of the phase-field model for the magnetic shape memory effect described in this work is the analysis of the interplay between magnetic domains and the motion of twin boundaries under an external applied magnetic field.[1] Lai et al. [170] have shown experimentally that under a moderate external magnetic field that favors one martensitic variant in a Ni_2MnGa sample, a complete alignment of moments with the fields direction occurs in the favored variant, but that the magnetic domain structure in the shrinking variant is almost uneffected. Shrinking of the domains within the unfavored variants only occurs due to the motion of the twin boundary (see [170, Figure 2] and Fig. 11.1), but no reorientation of the magnetic moments due to the external field or the arising demagnetization energy related to the head-to-head configuration is observed. A simulation scenario comparable to the the experimental one can be set-up to run simulation studies on the development of the domain structure in the unfavored martensitic variant. An interesting point would be the influence different initial magnetization configurations have on the development of both the martensitic variant and the magnetic domain structure. The results could be compared to results produced by other models, e.g. these developed by Kiefer and Lagoudas [158], Kiefer et al. [7] or Wang and Steinmann [6].

---

[1]The following discussion about this topic has been initiated by Prof. B. Kiefer of the University of Dortmund, Institute of Mechanics.

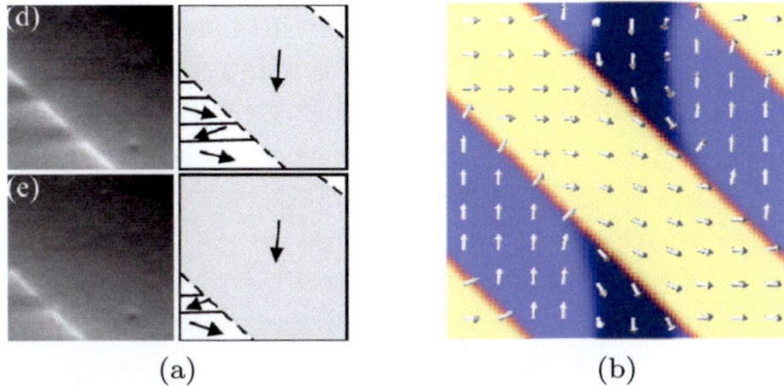

(a) (b)

Figure 11.1: Magnetic domain wall development in a Ni_2MnGa sampe, **(a)** observed experimentally (the figure is taken from Lai et al. [170]), and **(b)** observed in a simulation (cp. Fig. 10.4). Both show a magnetic domain wall structure in the unfavored martensitic variant that leads to unfavorable head-to-head configurations after the reorientation in the favored variant has occurred.

11.2 Shape memory materials and equilibrium elasticity

In first applications exceeding the investigations presented in this work, purely elastic simulations of strain accomodation in nano-grains of the conventional shape memory alloy NiTi under isothermal conditions have been carried out, motivated by discussions with Prof. Waitz[2] and the work of Waitz et al. presented in [171]. The size of the nano-grains does not exceed 150 nm, so that simulations could in principle deal with a whole grain, waiving the need for representative volumina. NiTi undergoes a cubic-to-monoclinic MT, giving rise to the formation of 12 different martensitic variants.[3] Thus, more complex microstructures than in Ni_2MnGa have

---

[2]Ao. Univ.-Prof. Mag. Dr. Thomas Waitz is professor at the department of physics, university of Vienna. The said discussions were held at the International Conference on Ferromagnetic Shape Memory Alloys (ICFSMA 2012) in Dresden in July 2011.

[3]Recall that the MT is symmetry breaking and that the cubic point group has 24, the monoclinic has two rotations symmetries (see e.g. [10] or [17]). So Thm. 2.2 gives 24 different variants. Six pairs of these can proven to be compatible in the sense of Def. 5.2 [10].

can possibly develop. The first results in the context of analyzing structures of compatible twins in NiTi are promising and have been presented by the author at the DPG Spring Meeting[4] in March 2012 in Berlin. The SOR-algorithm that is used to compute the mechanical equilibrium has proven well in many cases (see Secs. 7.3 and 10.3). Nevertheless, in some cases the SOR-algorithm tends to converge very slowly, what significantly limits the range of application and the feasible domain sizes. These cases need a better understanding. Numerical analysis concerning the condition of the problem and the applicability of the solution method has to be carried out properly. If necessary, the solution method has to be refined or replaced by a more accurate one. Having an efficient method at hand, the goal of simulation-assisted investigations of polycrystals under mechanical load is achievable. This opens up many interesting applications in different fields, e.g. the growth of polycrystals on thin films where the film and the substrate interact mechanically, or crack-sealing processes in geological sciences where mechanical processes play a major role in understanding crystal growth processes in the interior of the earth.

11.3 Magnetic thin films

Another application in the range of interest is the numerical analysis of magnetic thin films. Materials as the magnetic shape memory material Ni_2MnGa consist of different variants but have homogeneous magnetic exchange properties. If phases with different magnetic exchange properties come into play, compatibility conditions at their interfaces have to be maintained (for the so called *exchange coupling*, see the book of Hubert and Schäfer [44]). To make further use of the concept of RVEs, periodicity has to be insured in less than three dimensions (in opposition to the approach discussed in Sec. 8.2.2). In the literature, there exist proposals how periodic 1D and 2D boundary conditions for the magnetization can be realized (see the works of Lebecki et al. [114] or Wang et al. [115]). Their implementation might permit the realization of periodic RVEs of an infinitely long rod in 1D or an infinitely extended thin film in 2D. The main difficulty in extending a finite specimen to infinity

---

[4]The DPG Spring Meeting is the annual conference of the Deutsche Physkalische Gesellschaft e.V. See **www.dpg-physik.de/** for more information.

arises from the need of calculating the demagnetization field. The interactions of infinitely distant magnetic dipols have to be taken into account. The basic idea of the approaches cited above exploit the assumed periodicity of the calculation domain. They start from the demagnetization tensor calculation for a finite extended specimen (as in Chap. 8.2), expand one or two dimension to infinity and analyze how this affects the demagnetization tensor components. The exact formulae are used for a finite number of 'copies' of the reference computation domain (the unit of repetition), while for magnetic dipols far away integral approximations are used, for which analytic solutions exist. Mathematical and numerical analysis are necessary to verify the correctness of the expansion as well as the accuracy of the implementation. Care has to be taken when the exact interaction formulae derived by Newell at al. are evaluated (cp. [116] and Sec. 8.2.1): For large distances, as pointed out by Wang et al. in [115], this formulae result in numerical inaccuracy. As emphasized by Michael Donahue [172], care has to be taken generally when the demagnetization tensor components are computed, and thorough analysis of the equations and equivalent mathematical reformulation of the expressions can lead to a more stable numerical results. Periodic boundary conditions in 1D and 2D are available as extensions to the software framework OOMMF[5], and a first implementation could be based on the OOMMF-routines.

11.4 Ferroic cooling

The last field of research to be mentioned here, where the model and the methods shown in this work might be applied, is the at the moment highly investigated field of *ferroic cooling*, that is cooling based on the *magnetocaloric effect* (MCE). This effect was first discovered by Warburg in 1881 in iron (see [173]): The process of adiabatic demagnetization by on isothermally applying a magnetic field is based on the reduction of the configurational entropy, which can be exchanged in form of heat between the spin system and the crystal lattice. A following adiabatic removal of the field will cause a cooling of the sample. The gradual change of magnetization at the paramagnetic-ferromagnetic transition is related to a small

---

[5]Cp. Chap. 8.2 and `http://math.nist.gov/oommf/contrib/oxsext/`, the OOMMF extension modules website.

MCE and hence, large changes in magnetization due to a magnetic transition at the constant first order transition temperature will produce a large effect and thus a greater cooling power. Similar cooling principles hold for the application of an external stress field, resulting in the barocaloric or elastocaloric effect. The basic cooling principle is shown in Fig. 11.2 that is taken from [174].

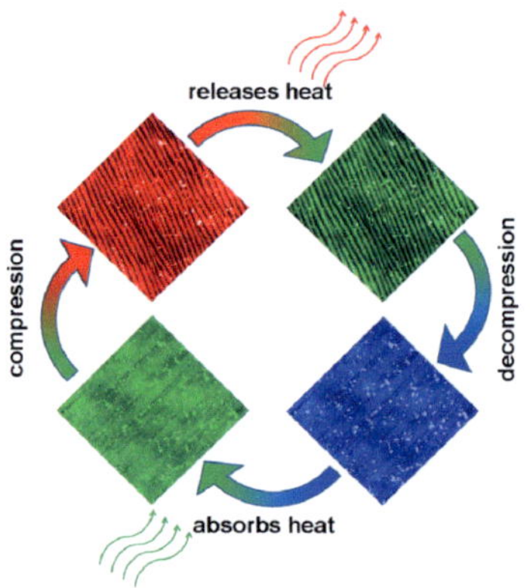

Figure 11.2: Illustration of the barocaloric effect taken from the article by Fähler et al. [174]: Starting with a material in the austenite state (lower left corner), a MT is induced by adiabatic application of external stress. A twinned microstructure develops, and the entropy decreases, so the temperature increases. The heat is then adsorbed, and the adiabatic pressure release lets the sample transform back into the austenite state and cools the sample further down.

In the last decade, giant entropy changes have been discovered in materials undergoing diffusionless first order and second order transitions at the same time, e.g. as in $Gd_5(Si_x, Ge_{1-x})_4$, where synchronously a first-order structural and magnetic transition appears [175]. An important class showing a large MCE are MSMAs, which may exhibit giant magneto-, baro- and elastocaloric effects [176]. Another group of alloys exhibit inverse (or negative) magnetocaloric effects, often attributed to a antiferromagnetic/ferromagnetic or antiferromagnetic/ferrimagnetic transition: The application of an external magnetic field causes the material to cool down. Materials with the highest potential are found within the class of ferromagnetic Heusler alloys based on the composition Ni_2MnX (with X as a third component), where a MT from a cubic austenite phase to a lower symmetric martensite phase is involved. Examples are $Ni_2Mn_{1+x}In_{1-x}$, for which a barocaloric effect of 24.4 $\frac{J}{KgK}$ close to room temperature has been reported [176], accompanied by an inverse magnetocaloric effect, or $Ni_2Mn_{1-x}Sn_{1+x}$ [177]. A system well-studied experimentally is the Heusler alloy Ni_2MnGa (cp. Sec. 5.1 and the simulations in Chap. 10). Ni_2MnGa shows a large caloric effect on application of a field near the transition temperature [178] and additionally the MSME. The entropy change induced by the external magnetic field $\mathbf{H}_{ext}$, $\langle S(\mathbf{H}_{ext})\rangle$, and the difference in magnetization between parent and martensitic phase strongly depends on the external field. It can be quantified in terms of micromagnetic and elastic energy contributions and the mesostructure [178]. Reviews on the MCE are given by Pecharsky and Gschneidner [179] and Gschneidner et al. [180, 181]. The theory of the MCE and the related thermodynamics are described by Oliveira and Ranke [182]. The martensitic transition in shape memory alloys plays an essential role, as it is a first order phase transition releasing latent heat and could be utilized for highly effective solid state cooling devices when processed in a cyclic mode. Momentarily, the research concentrates on the most preferable material systems showing simultaneously crystallographic and magnetic transitions. To conduct this screening systematically, understanding the physical mechanisms on which the interdependency of elastic and magnetization fields (magneto-elastic coupling) is based, is a necessary prerequisite. The processes have to be understood on the microscale as well as the mesoscopic length scale. An entropy-based formulation of the phase-field model and the coupling of a heat diffusion equation as proposed in [183] enables to account for temperature effects. As the saturation magnetization depends on the temperature, too, this dependency has to be included into the evolution equation for the spontaneous magnetization. A possibility is the implementation of the Landau-Lifshitz-Bloch equation (see [68]), that could substitute Eq. (4.8) to take thermal fluctuations into account. With that and the developed methods discussed in this work, the phase-field model might provide useful tools to support simulation-based microstructure investigations. These may provide interesting insights towards an understanding of the mesoscopic processes that lead to a giant (inverse) MCE, and thus support the development of enhanced environment-friendly cooling devices.

Part V

Appendix

A Interpretation and representation of rotations

Rotations were introduced as special linear isometries in Sec. 2.3. There are several ways to represent rotations in Euclidean spaces. The most common representation is the representation by orthogonal matrices. But for some purposes, if many rotations are to be concatenated, the representation by so called unit quaternions might be considerable. The chapter briefly discusses some general aspects about the interpretation of rotations, and the two different representations of rotations as orthogonal matrices and unit quaternions. Additionally, Euler angle conventions are briefly discussed, as this is the way rotations are used in the software Pace3D, which was used for all simulations in this work. The explanations of the following sections make extensive use of the notation introduced in Chap. 2.

A.1 Orthogonal matrices and interpretation of rotations

Rotations in n-space can be represented by orthogonal matrices with unit determinant, i.e. elements of the matrix group $SO(n)$. In this section, the case of $n = 3$ is considered, and the vector space $\mathbb{R}^3$ is thought to be equipped with standard scalar product '$\cdot$'. To give a meaning to vectors as 3-tuples of real-valued numbers, a frame of reference $\{O, \mathbf{x}, \mathbf{y}, \mathbf{z}\}$ is fixed at $O = \mathbf{0} \in \mathbb{R}^3$. Reasoning in this section follows in many aspects the books of Newnham [55] and Goldstein [184].

Convention To simplify the discussion in this section, $O \in \mathbb{R}^3$ is fixed as a common origin of all occurring frames of reference in this section. So, two different frames of reference only differ by a rotation, but not by a non-trivial translation. As an abbreviation for two frames of reference $(O, \mathbf{x} = \mathbf{x}_1, \mathbf{y} = \mathbf{x}_2, \mathbf{z} = \mathbf{x}_3)$ and $(O, \mathbf{x}' = \mathbf{x}_1', \mathbf{y}' = \mathbf{x}_2', \mathbf{z}' = \mathbf{x}_3')$ only O and O' will be used.

The section starts with some general considerations about the interpretation of the entries of rotation matrices and vectors. A rotation matrix acts as follows: Assume two orthonormal coordinate systems that share the same origin: An 'old' one denoted by O, and a 'new' one denoted by O'. Project the i-th axis $\mathbf{x}_i'$ of the new system onto the j-th axis $\mathbf{x}_j$ of the old system. As the systems are orthonormal, it is given by the *direction cosine*, i.e. the cosine of the angle between $\mathbf{x}_i'$ and $\mathbf{x}_j$:

$$r_{ij} := \cos(\angle(\mathbf{x}_i', \mathbf{x}_j)) = \mathbf{x}_i' \cdot \mathbf{x}_j.$$

There are nine direction cosines r_{ij}, and from the definition of angles this describes the rotation $R = (r_{ij})$ needed to rotate $\mathbf{x}_j$ onto $\mathbf{x}_i'$ in an *anti-clockwise sense*, or to rotate the system O onto the system O' in an *anti-clockwise sense*.

Let $v = (x, y, z)^T \in \mathbb{R}^3$ be a vector with components with respect to O. By projecting this vector onto the new axes $\mathbf{x}', \mathbf{y}', \mathbf{z}'$, one gets the components of v with respect to O'. In this interpretation v has not been moved in space. But by taking the relation of an anti-clockwise rotation from O onto O' into account, the new components of v can be interpreted in the old system O. Then R has moved v by a *clockwise* rotation in space.

Remark The definition of a rotation matrix $R = (r_{ij})$ can be easily altered by changing the definition of the direction cosines to $r_{ij} := \cos(\angle(\mathbf{x}_i, \mathbf{x}_j'))$, such that the orientation relation is inverted to a *clockwise rotation*. This arguments shows that interpretation and sense of a rotation are not fixed, but have to be well defined to give sense to the interpretation of a rotation.

So, when talking about a rotation $R \in SO(3)$, meaning has to be given to the action of R on $\mathbb{R}^3$. There are basically two things to fix: The sense of the rotation (*clockwise* or *anti-clockwise*), and if the action is defined with respect to the movement of vectors in a fixed reference system or changes the coordinates of a vector by changing the frame of reference (what leaves every point in the system 'in its originals place'). The following convention is applied here.

Convention A coordinate change by rotation of a frame of reference happens *anti-clockwise*, and (consequently) vectors are moved in space *clockwise*.

So, rotations either change the frame of reference, or move a vector in space. This is just a matter of the point of view, what motivates the following definition:

Definition A.1 (ACTIVE AND PASSIVE VIEW) Let $R \in \mathrm{SO}(3)$ be a rotation in the frame of reference O. The action of R on $\mathbb{R}^3$ can be interpreted in two different ways with respect to the action of the vector in $\mathbb{R}^3$. Let $v \in \mathbb{R}^3$ be a vector.

Active view R changes the position of v in $\mathbb{R}^3$, i.e. the vector

$$v' = Rv$$

is described in the same frame of reference as v, but may have another position than v (when v is not on the axis of rotation), i.e. $v' \underset{\text{in general}}{\neq} v$.

Passive view R does not change the position of v in $\mathbb{R}^3$, i.e. the vector

$$(v)' = Rv$$

has the same position in space, but its coordinates are described in another frame of reference as v, namely O' which has the relation R to the former.

$\square$

Remark In the *active view*, one can think of vectors to be moved around in space, whereas the *passive view* describes the change of coordinates (that is a *basis transformation* in the sense of Def. 2.9).

Let $R \in \mathrm{SO}(3)$ be a rotation, and let $v \in \mathbb{S}^2$ be an eigenvector to the eigenvalue 1 (cp. Thm. 2.9). Then R can be represented as the rotation around v about an angle $\alpha \in [0, 2\pi[$, written $R = R(v, \alpha)$.

If rotations are concatenated, they may refer to a fixed immobile set of axes in a fixed frame of reference, or they might refer to the new axes one gets after the rotations are carried out successively. As Euler angles will be used to describe the orientations between two frames of reference, the following definition is restricted to the case of three successive rotations.

Definition A.2 (INTRINSIC AND EXTRINSIC VIEW) Let $e_\alpha, e_\beta, e_\gamma \in \mathbb{S}^2$ be three (not necessarily different) fixed global axes of the frame of reference O, and let $(\alpha, \beta, \gamma) \in [0, 2\pi[^3$ be the angles of rotations around these three axes. There are two ways to interpret the rotation induced by these angles, dependent on whether the axes of the global frame of are thought to stay fixed or not:

Intrinsic view (α, β, γ) are interpreted as rotations around the axis that are generated by the former rotations. So the rotations carried out are $R(e_\alpha, \alpha)$, $R(R(\alpha, e_\alpha)e_\beta, \beta)$ and $R(R(\beta, e_\beta)R(\alpha, e_\alpha)e_\gamma, \gamma)$.

Extrinsic view (α, β, γ) are interpreted as rotations around the axes of the global frame of reference. So carried out are $R(e_\alpha, \alpha)$, $R(e_\beta, \beta)$ and $R(e_\gamma, \gamma)$.

$\square$

Remark So, in the *intrinsic view* rotations are applied around axes that change during the process of rotation, while in the *extrinsic view* the rotations act on three fixed global coordinate axes.

The semantics of these transformations have to carefully defined, as the algebra of matrices does not know intentions. According to the above definitions, there are four possible combinations:

1. Extrinsic active view.

2. Extrinsic passive view.

3. Intrinsic active view.

4. Intrinsic passive view.

When changing from one frame of reference O to another frame of reference O', the action of any operator $A \in \mathrm{GL}(n)$ defined in O can be described in O': As the action of A is known in O, as well as the relation R between O and O', one can go back from O' to O using $R^{-1} = R^T$, apply A in O, and return to O' using R. This is the concept of similarity transformations, and the result reads $A' = RAR^T$.

A.2 Basic rotation matrices and Euler angles

Now, the rotation matrices for rotations around the standard basis in $\mathbb{R}^3$ for the active and passive view are given. To fix semantics, the active view refers to the anti-clockwise rotation of vectors, and the the passive view refers to the anti-clockwise rotation of coordinate frames. The Fig. A.1 illustrates the definition.

Definition A.3 (ROTATION MATRICES AROUND THE GLOBAL AXES) As rotation matrices for the active and passive view are to be defined, the active view is indicated by the superscript a and the passive view with the superscript p. Let $\varphi \in [0, 2\pi[$ be an angle of rotation.

Rotations around the x-axis

$$R^a(\mathbf{x}, \varphi) = \begin{pmatrix} 1 & 0 & 0 \\ 0 & \cos\varphi & -\sin\varphi \\ 0 & \sin\varphi & \cos\varphi \end{pmatrix} \quad R^p(\mathbf{x}, \varphi) = \begin{pmatrix} 1 & 0 & 0 \\ 0 & \cos\varphi & \sin\varphi \\ 0 & -\sin\varphi & \cos\varphi \end{pmatrix}$$

Rotations around the y-axis

$$R^a(\mathbf{y}, \varphi) = \begin{pmatrix} \cos\varphi & 0 & \sin\varphi \\ 0 & 1 & 0 \\ -\sin\varphi & 0 & \cos\varphi \end{pmatrix} \quad R^p(\mathbf{y}, \varphi) = \begin{pmatrix} \cos\varphi & 0 & -\sin\varphi \\ 0 & 1 & 0 \\ \sin\varphi & 0 & \cos\varphi \end{pmatrix}$$

Rotations around the z-axis

$$R^a(\mathbf{z}, \varphi) = \begin{pmatrix} \cos\varphi & -\sin\varphi & 0 \\ \sin\varphi & \cos\varphi & 0 \\ 0 & 0 & 1 \end{pmatrix} \quad R^p(\mathbf{z}, \varphi) = \begin{pmatrix} \cos\varphi & \sin\varphi & 0 \\ -\sin\varphi & \cos\varphi & 0 \\ 0 & 0 & 1 \end{pmatrix}$$

$\square$

(a)   (b)

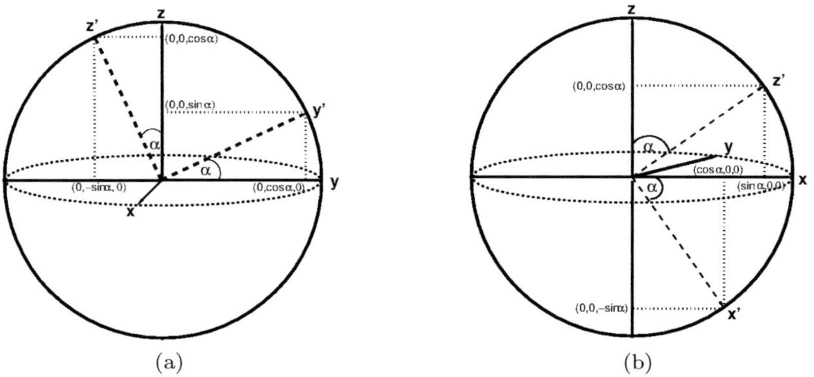

Figure A.1: Sketch of the rotations around one of the standard axes about the angle α, interpreted as rotating the original coordinate frame anticlockwise (passive view): **(a)** Rotation around the **x**-axis, and **(b)** rotation around the **y**-axis. The rotation around **z** follows from the one around **x** by renaming the axes. The derivation of the rotation matrices follows directly from the drawings by projecting the 'new' axes onto the 'old' ones.

From Def. A.3 immediately follows the relation between matrices representing the active and the passive view.

Remark The relation between the active and the passive view is given by matrix transposition, i.e. for all axes $\mathbf{a} \in \{\mathbf{x}, \mathbf{y}, \mathbf{z}\}$ and angles $\varphi \in [0, 2\pi[$ the relation

$$R^p(\mathbf{a}, \varphi) = (R^a(\mathbf{a}, \varphi))^T$$

holds. This relation is also valid in the general case when rotations around arbitrary axes $\mathbf{a} \in \mathbb{S}^2$ are allowed.

Interpretation Let $\mathbf{a} \in \{\mathbf{x}, \mathbf{y}, \mathbf{z}\}$ and $\varphi \in [0, 2\pi[$.

- The *active rotation* $R^a(\mathbf{a}, \varphi)$ can either be interpreted as moving a vector *anti-clockwise* around $\mathbf{a}$, or transforming the frame of reference *clockwise*.

- The *passive rotation* $R^p(\mathbf{a}, \varphi)$ can either be interpreted as transforming the frame of reference *anti-clockwise*, or moving a vector *clockwise* around $\mathbf{a}$.

Each rotation R can be (non-uniquely) decomposed into three successive rotations about three angles around three defined orthonormal axes, because the degree of freedom a rotation has is three (cp. [184, Chap. 4.1]). Hence, every rotation is completely determined by a triple $(\alpha, \beta, \gamma) \in [0, 2\pi[^3$. This is used to define the so called Euler angles (cp. [184]).

Definition A.4 (EULER ANGLES) Let O be a frame of reference and $(\alpha, \beta, \gamma) \in [0, 2\pi[^3$ be angles of rotations. The triple (α, β, γ) are the *Euler angles* of the rotation arising by rotating about α, then β and finally γ around three orthogonal axes, such that no two successive rotations are around the same axis. $\qquad\qquad\square$

The definition of Euler angles has advantages as well as disadvantages. As there are no axes or orders fixed, different conventions are possible.

Remark

- There are twelve possible Euler angles conventions.

- There are six possible Euler angles convention when all axes of rotation have to be different.

- An interpretation has to be given to the resulting rotation: active or passive, intrinsic or extrinsic.

- Different rotations may lead to the same result.

- A rotation may move one axis parallel to another one, what leads to a loss of a degree of freedom. This is called a *gimbal lock*.

In the following one interpretation of a given triple of Euler angles (α, β, γ) is discussed for the frame of reference $\{O, \mathbf{x}, \mathbf{y}, \mathbf{z}\}$. Assume the change coordinates anti-clockwise by applying rotations

1. first around the $\mathbf{x}$-axis,

2. then around the $\mathbf{y}$-axis,

3. then around the $\mathbf{z}$-axis.

This fixes the interpretation as the extrinsic passive view. Remember that a rotation affects the whole space and coordinates shall be transformed. When the rotations are applied around $\mathbf{x}$, $\mathbf{y}$and $\mathbf{z}$, these axes are affected by the transformations. The concept of similarity transformation allows to go back to the 'original' frame of reference. Let (α, β, γ) be an Euler angle triple. The first rotation is

$$R_1 := R^p(\mathbf{x}, \alpha).$$

The next rotation shall be around the fixed 'old' $\mathbf{y}$-axis, hence the second rotation has to be applied in the original system O:

$$R^p(\mathbf{x}, \alpha) R^p(\mathbf{y}, \beta) (R^p(\mathbf{x}, \alpha))^T,$$

what applied to $R^p(\mathbf{x}, \alpha)$ results in

$$R_2 := R^p(\mathbf{x}, \alpha) R^p(\mathbf{y}, \beta)$$

for the rotation around $\mathbf{y}$. The same concept is used for the last rotation around the 'old' $\mathbf{z}$-axis:

$$R_3 := (R^p(\mathbf{x}, \alpha) R^p(\mathbf{y}, \beta)) R^p(\mathbf{z}, \gamma) (R^p(\mathbf{x}, \alpha) R^p(\mathbf{y}, \beta))^T$$

applied to $R^p(\mathbf{x}, \alpha) R^p(\mathbf{y}, \beta)$ gives

$$R^p(\mathbf{x}, \alpha) R^p(\mathbf{y}, \beta) R^p(\mathbf{z}, \gamma),$$

such that

$$R_3 R_2 R_1 = R^p_{\mathbf{x}, \mathbf{y}, \mathbf{z}} = R^p(\mathbf{x}, \alpha) R^p(\mathbf{y}, \beta) R^p(\mathbf{z}, \gamma)$$

is the wanted rotation matrix. When the matrix is transposed, a change to the extrinsic active view is made, in which vectors are moved anti-clockwise around the fixed axes $\mathbf{x}$, $\mathbf{y}$ and $\mathbf{z}$ as

$$R^a_{\mathbf{x},\mathbf{y},\mathbf{z}} = (R^p_{\mathbf{x},\mathbf{y},\mathbf{z}})^T = R^a(\mathbf{z},\gamma)R^a(\mathbf{y},\beta)R^a(\mathbf{x},\alpha).$$

In the Pace3D solver environment the extrinsic active view is assumed that moves vectors anti-clockwise around fixed $\mathbf{x}$-, $\mathbf{y}$- and $\mathbf{z}$-axes, and changes coordinate systems clockwise around the three fixed axes $\mathbf{z}$, then $\mathbf{y}$, then $\mathbf{x}$. The rotation matrix representing the movement of a vector reads

$$R^a_{\mathbf{x},\mathbf{y},\mathbf{z}} = R^a(\mathbf{z},\gamma)R^a(\mathbf{y},\beta)R^a(\mathbf{x},\alpha), \tag{A.1}$$

and when interpreting this matrix as an extrinsic passive view that changes coordinate frames, is rewritten in terms of matrices in the passive view as

$$R^a_{\mathbf{x},\mathbf{y},\mathbf{z}} = (R^p(\mathbf{z},\gamma))^T (R^p(\mathbf{y},\beta))^T (R^p(\mathbf{x},\alpha))^T. \tag{A.2}$$

A.3 Interpretation of phase orientations

Sometimes the Euler angles are interpreted in the way of Eq. (A.1) as extrinsic active view, and sometimes the exact same matrix is interpreted it as the orientation relationship between coordinate systems in the sense of Eq. (A.2) as an extrinsic passive view. This is the case e.g. when phases with different elastic properties are considered, reflected by different elastic property tensors $\mathcal{C}$. This tensor is usually described in the coordinate system associated with a phase p, where $\mathcal{C} = \mathcal{C}_p$ has a simpler form as the material symmetries p provides are reflected. Because the system consists of different phases, energies and properties like stresses and strains are evaluated in a common coordinate system O_Ω associated to the simulation box. When $\mathcal{C}_p$ is described in O_Ω, denoted $\mathcal{C}_\Omega$, the 'nice form' of $\mathcal{C}_p$ is lost during the change of coordinates. $\mathcal{C}_\Omega$ is gained from the relation between O_Ω and O_p, that is described by the Euler angles according to Eq. (A.2) by successive clockwise rotations of the coordinate axes around the fixed axes $\mathbf{z}$, $\mathbf{y}$ and $\mathbf{x}$, resulting in a rotation R. To get $\mathcal{C}_\Omega$ from $\mathcal{C}_p$, R^T

has to be applied. Figure A.2 illustrates the situation: The translation T between the two origins can be omitted.

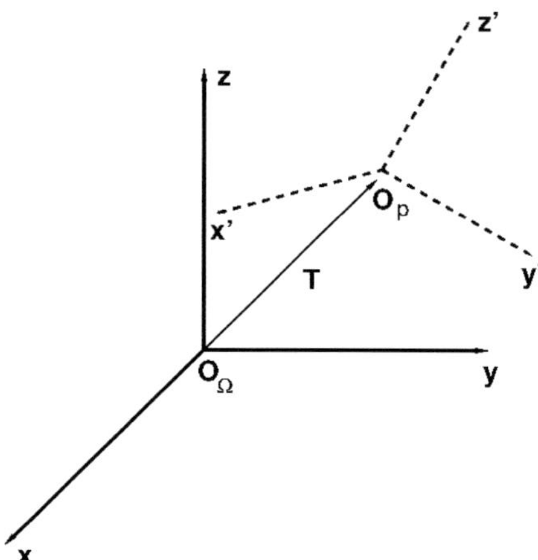

Figure A.2: Relation between two coordinate systems: The frame of reference O_Ω and the coordinate system of phase p. The translation T is shown to make the sketch easier understandable.

A.4 Quaternions

This section gives a very brief survey of the usage of unit quaternions to represent rotations in the Euclidean three-space. This overview contains two parts: A more theoretical part to legitimatize the idea to use quaternions as representations of rotations, and a second part, in which the transformation formulae are given explicitly. The writing of this text is rather informal, theorems are stated implicitly and no proofs are given. Rigorous definitions and analysis can be read in the books of Beutelspacher [185], Stoth [186] and Selig [187].

$\mathbb{C}$, the field of complex numbers, is the only[1] non-trivial field extension of $\mathbb{R}$ with finite index, i.e. there exists no field $\mathbb{R} \subset \mathbb{K}$ such that $|\mathbb{K} : \mathbb{R}| < \infty$

---

[1]except for isomorphisms

but $\mathbb{K} = \mathbb{C}$.[2] In 1843, *Hamilton* discovered a set of operations which 'almost' make the vector space $\mathbb{R}^4$ to an extension field of the reals, but that lack of commutativity.

Let

$$\mathbb{H} := \{(q_0, q_1, q_2, q_3)^T \mid q_0, q_1, q_2, q_3 \in \mathbb{R}\} = \mathbb{R}^4$$

be the set of all quadrupels with real components. For all $q, q' \in \mathbb{H}$ with $q = (q_0, q_1, q_2, q_3)^T$ and $q' = (q'_0, q'_1, q'_2, q'_3)^T$ define a (component-wise) *addition* by

$$q + q' := (q_0 + q'_0, q_1 + q'_1, q_2 + q'_2, q_3 + q'_3)^T$$

and a *multiplication* by

$$\begin{aligned} qq' := q \cdot q' := (&q_0 q'_0 - q_1 q'_1 - q_2 q'_2 - q_3 q'_3, \\ &q_0 q'_1 + q_1 q'_0 + q_2 q'_3 - q_3 q'_2, \\ &q_0 q'_2 + q_2 q'_0 + q_3 q'_1 - q_1 q'_3, \\ &q_0 q'_3 + q_3 q'_0 + q_1 q'_2 - q_2 q'_1)^T. \end{aligned}$$

With these operations, $\mathbb{H}$ becomes a *skew field* (cp. Def. 2.7) with the following properties

1. $(\mathbb{H}, +)$ is an abelian group with $\mathbf{0} := (0, 0, 0, 0)^T$ as *zero*. The (additive) *inverse* to $q = (q_0, q_1, q_2, q_3)^T \in \mathbb{H}$
   is $-q = (-q_0, -q_1, -q_2, -q_3)^T$.

2. $(\mathbb{H} \setminus \{\mathbf{0}\}, \cdot)$ is a non-abelian group with $\mathbf{1} := (1, 0, 0, 0)^T$ as *one*. The (multiplicative) *inverse* to $q = (q_0, q_1, q_2, q_3)^T \in \mathbb{H} \setminus \{\mathbf{0}\}$ is $q^{-1} = (\frac{q_0}{|q|}, \frac{q_1}{|q|}, \frac{q_2}{|q|}, \frac{q_3}{|q|})^T$, where $|q|$ is the Euclidean length of q.

3. The law of *distributivity* is valid:

$$q \cdot (q' + q'') = qq' + qq'' \text{ for all } q, q', q'' \in \mathbb{H}.$$

$\mathbb{R}$ can be embedded[3] into $\mathbb{H}$ via the map

$$\tilde{} : \mathbb{R} \to \mathbb{H}, x \mapsto \tilde{x} := (x, 0, 0, 0).$$

---

[2] $|\mathbb{K} : \mathbb{R}|$ denotes the *index* of $\mathbb{K}$ over $\mathbb{R}$ and is the dimension of $\mathbb{K}$ considered as an $\mathbb{R}$-vectors pace.

[3] Embeddings are injective maps, and therefore invertible.

In analogy to $\mathbb{C}$, three *imaginary units* are defined by

$$\mathbf{i} := (0,1,0,0)^T, \; \mathbf{j} := (0,0,1,0)^T \; \text{ and } \; \mathbf{k} := (0,0,0,1)^T.$$

For these

$$\widetilde{-1} = \mathbf{ii} = \mathbf{jj} = \mathbf{kk} = \mathbf{ijk}$$

and

$$\mathbf{ij} = \mathbf{k}, \quad \mathbf{jk} = \mathbf{i}, \quad \mathbf{ki} = \mathbf{j}$$

hold. An example for the non-commutativity of $\mathbb{H}$ shows

$$\mathbf{ij} = \mathbf{k} \neq -\mathbf{k} = \mathbf{ji}.$$

Any quaternion $q = (q_0, q_1, q_2, q_3)^T \in \mathbb{H}$ can be uniquely written in terms of the imaginary units as

$$q = \widetilde{q}_0 + q_1\mathbf{i} + q_2\mathbf{j} + q_3\mathbf{k}.$$

Hence $|\mathbb{H} : \mathbb{R}| = 4$.

Let $q = (q_0, q_1, q_2, q_3)^T \in \mathbb{H}$ be a quaternion. $\Re(q) = q_0$ is the *real part* (or *scalar part*) of q, and $\Im(q) := (q_1, q_2, q_3)^T$ is the *imaginary part* (or *vector part*) of q. With

$$\Im(\mathbb{H}) := \{(q_0, q_1, q_2, q_3)^T \in \mathbb{H} \mid q_0 = 0\}$$

the set of pure imaginary quaternions is denoted. The *conjugated quaternion* to q is $\bar{q} := (q_0, -q_1, -q_2, -q_3)$. If $|q| = 1$, then q is a *unit quaternion*. The set of all *unit quaternions*

$$\mathbb{S}^3 := \{q \in \mathbb{H} \mid |q| = 1\}$$

is a three dimensional subspace of $\mathbb{H}$. The relation between inversion and conjugation is given by the length of the quaternion q (in analogy to $\mathbb{C}$):

$$q^{-1} = \frac{1}{|q|^2}\bar{q}.$$

Hence, $\mathbb{S}^3$ is a multiplicative subgroup of $(\mathbb{H} \setminus \{\mathbf{0}\}, \cdot)$ with quaternion conjugation as inversion.

Unit quaternions can be used to represent rotations in the Euclidean three-space. For this, the following observation is important:

The group $\mathbb{S}^3$ operates transitively (and non-freely) on the set $\Im(\mathbb{H})$ via conjugation, and the operation leaves distances and directions invariant, i.e. the operations represent rotations of $\Im(\mathbb{H})$. The *conjugation map* is defined as (cp. Def. 2.2)

$$q^s := sq\bar{s} \text{ for all } s \in \mathbb{S}^3 \text{ and } q \in \Im(\mathbb{H}).$$

The map

$$\varphi : \mathbb{R}^3 \to \Im(\mathbb{H}), (x_1, x_2, x_3)^T \mapsto (0, x_1, x_2, x_3)^T$$

is the natural embedding of the Euclidean three-space $\mathbb{R}^3$ into $\mathbb{H}$, so $\mathbb{R}^3$ can be identified with the imaginary part of the quaternion skew field. Hence, unit quaternions can be used to represent rotations of vectors of $\mathbb{R}^3$. The rotation represented by the unit quaternion $s \in \mathbb{S}^3$ is given by

$$\Im(\mathbb{H}) \to \Im(\mathbb{H}), h \mapsto h^s = sh\bar{s},$$

and therefore the rotation action of $s \in \mathbb{S}^3$ on $\mathbb{R}^3$ is given by

$$\rho_s : \mathbb{R}^3 \to \mathbb{R}^3, x \mapsto \varphi^{-1}(\varphi(x)^s) = \varphi^{-1}(s\varphi(x)\bar{s}).$$

The following properties hold:

- For all $s, s' \in \mathbb{S}^3$: $\rho_s \circ \rho_{s'} = \rho_{ss'}$, i.e. the concatenation of rotations is represented by quaternion multiplication.
- The unit quaternions s and $-s$ represent the same rotation.
- For $s \in \mathbb{S}^3$ the inverse rotation is represented by $\bar{s}$.

Interpretation Let $q \in \mathbb{S}^3$ be a unit quaternion. q represents the rotation of a certain angle around a certain axis in three-space. In fact, the components of q can be interpreted as follows:
The real part of q is the cosine of twice the rotation angle, the imaginary part gives the axis of rotation, compressed by the sine of twice the rotation angle.

Let $q = (q_0, q_1, q_2, q_3)^T \in \mathbb{S}^3$ be a unit quaternion. Then the rotation angle α and the axis of rotation $v \in \mathbb{S}^2$ can be computed following the above interpretation. As

$$q = \cos(2\alpha) + \sin(2\alpha)\widetilde{\Im(q)}.$$

one gets

$$\alpha = \frac{1}{2}\cos^{-1} q_0 \quad \text{and} \quad v = (v_1, v_2, v_3)^T \quad \text{with} \quad v_i = \frac{1}{2\alpha}\sin^{-1} q_i, \quad i = 1, 2, 3.$$

Let $\alpha \in [0, 2\pi[$ be an angle, and $v = (v_1, v_2, v_3)^T \in \mathbb{S}^2$ an axis of rotation in three-space. The the unit quaternion

$$q := \left(\cos\left(\frac{1}{2}\alpha\right), \sin\left(\frac{1}{2}\alpha\right) v_1, \sin\left(\frac{1}{2}\alpha\right) v_2, \sin\left(\frac{1}{2}\alpha\right) v_3 \right)$$

then represents the rotation around v by the angle α. In analogy to the rotation matrices introduced in Sec. A.2, the unit quaternions representing rotations of an angle $\varphi \in [0, 2\pi[$ around the standard unit axes of a Euclidean coordinate system $\{O, \mathbf{x}, \mathbf{y}, \mathbf{z}\}$ can be written explicitly:

- Rotation around the **x-axis**: $q_x = \left(\cos\left(\frac{1}{2}\alpha\right), \sin\left(\frac{1}{2}\alpha\right), 0, 0\right).$
- Rotation around the **y-axis**: $q_y = \left(\cos\left(\frac{1}{2}\alpha\right), 0, \sin\left(\frac{1}{2}\alpha\right), 0\right).$
- Rotation around the **z-axis**: $q_z = \left(\cos\left(\frac{1}{2}\alpha\right), 0, 0, \sin\left(\frac{1}{2}\alpha\right)\right).$

If now $(\varphi_x, \varphi_y, \varphi_z) \in \mathbb{R}^3$ is a triple of Euler angles rotations, the quaternion representing the rotation around the **x-**, **y-** **z-**axis in this order is given as

$$q_x q_y q_z.$$

Keeping the interpretations discussed in Sec. A.2 in mind, rotations can be constructed for any Euler angle convention.

The rest of this section shortly discusses how unit quaternions can be used to rotate vectors, the conversion rules between orthogonal matrices and unit quaternions and the comparison of computational costs of the implementation of rotations as orthogonal matrices and unit quaternions.

Convention For simplicity the application of the map φ that identifies $\mathbb{R}^3$ and $\Im(\mathbb{H})$ will be silently suppressed.

To rotate a vector $x = (x_1, x_2, x_3)^T \in \mathbb{R}^3$ by a rotation represented by a unit quaternion $q = (q_0, q_1, q_2, q_3)^T$ to get a vector x', one needs to embed

x into $\mathbb{H}$ and conjugate the result by q and go back to $\mathbb{R}^3$. Applying the above convention, this reads

$$x' = qx\bar{q}.$$

Because $x' \in \Im(\mathbb{H})$, it can be directly computed as

$$x' = (0, \tag{A.3}$$
$$2(x_1(-q_2^2 - q_3^2) + x_2(q_1q_2 - q_0q_3) + x_3(q_1q_3 + q_0q_2)) + x_1,$$
$$2(x_1(q_1q_2 + q_0q_3) + x_2(-q_1^2 - q_3^2) + x_3(q_2q_3 - q_0q_1)) + x_2,$$
$$2(x_1(q_1q_3 - q_0q_2) + x_2(q_2q_3 + q_0q_1) + x_3(-q_1^2 - q_2^2)) + x_3).$$

For implementation purposes, one can define nine temporary variables to speed up the computation:

$$\begin{aligned}
t_0 &= -q_1^2, & t_3 &= q_1q_2, & t_6 &= q_0q_2, \\
t_1 &= -q_2^2, & t_4 &= q_0q_3, & t_7 &= q_2q_3, \\
t_2 &= -q_3^2, & t_5 &= q_1q_3, & t_8 &= q_0q_1.
\end{aligned}$$

Then, the result $x' = (x_1', x_2', x_3')$ computes as

$$x_1' = 2(x_1(t_1 + t_2) + x_2(t_3 - t_4) + x_3(t_5 + t_6)) + x_1, \tag{A.4}$$
$$x_2' = 2(x_1(t_3 + t_4) + x_2(t_0 + t_2) + x_3(t_7 - t_8)) + x_2,$$
$$x_3' = 2(x_1(t_5 - t_6) + x_2(t_7 + t_8) + x_3(t_0 + t_1)) + x_3.$$

Let $x \in \mathbb{R}^3$ and $q \in \mathbb{S}^3$ be given, and let x' be the result of rotating x by the rotation represented by $q = (q_0, q_1, q_2, q_3)$. The 3×3 orthogonal matrix R representing the same rotation $x' = Rx$ can be constructed by rearranging the expressions in Eq. (A.3):

$$R = \begin{pmatrix}
2(-q_2^2 - q_3^2) + 1 & 2(q_1q_2 - q_0q_3) & 2(q_1q_3 + q_0q_2) \\
2(q_1q_2 + q_0q_3) & 2(-q_1^2 - q_3^2) + 1 & 2(q_2q_3 - q_0q_1) \\
2(q_1q_3 - q_0q_2) & 2(q_2q_3 + q_0q_1) & 2(-q_1^2 - q_2^2) + 1
\end{pmatrix}.$$

Now the amount of elementary algebraic operations[4] can be counted that are necessary to carry out rotations in different representations. Remind

---

[4] i.e. additions/subtractions and multiplications

that the matrix-vector and matrix-matrix multiplications to compute $v' = Rv$ and $R'' = RR'$ ($v = (v_1, v_2, v_3)^T$, $v' = (v'_1, v'_2, v'_3)^T \in \mathbb{R}^3$, $R, R' \in SO(3)$) are given by

$$v'_i = \sum_{j=1}^{3} R_{ij} v_j \qquad \text{and} \qquad R''_{ij} = \sum_{k=1}^{3} R_{ik} R'_{kj}, \qquad i, j = 1, 2, 3.$$

The following discussion refers to the implementation of formula Eq. (A.4). Then, by assuming the implementation by using a data type called 'REAL' for the numerical implementation,[5] simple counting leads to the memory and computation costs listed in the Tabs. A.1, A.2 and A.3. It can be seen that unit quaternions need less memory than matrices and less elementary operations when many rotation maps are concatenated (as it is the case in many visualization applications). When only vectors are transformed, the implementation of rotations as orthogonal matrices is favorable.

Table A.1: Memory usage for the implementation using entries of the type REAL.

	Total of REAL elements
Matrix	9
Quaternion	4

Table A.2: Number of operations for the rotation of a vector.

	Multiplications	Add./Sub.	Total of operations
Matrix	9	6	**15**
Quaternions	21	15	**36**

Table A.3: Number of operations for the concatenation of two rotations.

	Multiplications	Add./Sub.	Total of operations.
Matrix	27	18	**45**
Quaternions	16	12	**28**

---

[5]in the actual implementation this might be a float, a double or a long double, depending on the required accuracy.

B The Voigt notation for elasticity

To avoid computational expensive tensor operations, the so called *Voigt notation* is introduced here as a representation of Hooke's law using matrices and vectors. In this representation, computations can be carried out more efficiently, even when different crystallographically orientated phases are considered. This chapter derives the matrix-vector representation and gives the matrices for elastic stiffness tensors for often used crystal systems. Finally, a compact version of the update scheme for the displacement field $\mathbf{u}$ in the SOR algorithm presented in Sec. 7.3 is shown.

B.1 Matrix representation of Hooke's law

As mechanical stress $\boldsymbol{\sigma}$ and strain $\boldsymbol{\epsilon}$ are symmetric second rank tensors, they have only six independent entries each, and the so called Voigt notation can be used to represent these tensors as vectors of length six.

Definition B.1 (VOIGT REPRESENTATION OF ELASTIC STRESS, STRAIN AND THE ELASTIC STIFFNESS TENSOR) Let $\boldsymbol{\sigma} = (\sigma_{ij}), \boldsymbol{\epsilon} = (\epsilon_{ij}) \in \mathrm{symm}(\mathbb{R}^{3\times3})$ be tensors representing the elastic stress and elastic strain, respectively. Define an index map φ for all pairs $(i,j) \in \{1,2,3\}^2$ via

$$(i,j) \overset{\varphi}{\mapsto} \begin{cases} i, & \text{if } i = j \\ 9 - i - j, & \text{if } i \neq j \end{cases}.$$

The *Voigt representation* of $\boldsymbol{\sigma}$ and $\boldsymbol{\epsilon}$, denoted by $\boldsymbol{\sigma}_V$ and $\boldsymbol{\epsilon}_V$, is defined by

$$\boldsymbol{\sigma}_V = \begin{pmatrix} \sigma_1^V \\ \sigma_2^V \\ \sigma_3^V \\ \sigma_4^V \\ \sigma_5^V \\ \sigma_6^V \end{pmatrix} = \begin{pmatrix} \sigma_{11} \\ \sigma_{22} \\ \sigma_{33} \\ \sigma_{23} \\ \sigma_{13} \\ \sigma_{12} \end{pmatrix} \quad \text{and} \quad \boldsymbol{\epsilon}_V = \begin{pmatrix} \epsilon_1^V \\ \epsilon_2^V \\ \epsilon_3^V \\ \epsilon_4^V \\ \epsilon_5^V \\ \epsilon_6^V \end{pmatrix} = \begin{pmatrix} \epsilon_{11} \\ \epsilon_{22} \\ \epsilon_{33} \\ 2\epsilon_{23} \\ 2\epsilon_{13} \\ 2\epsilon_{12} \end{pmatrix}.$$

The fourth rank elastic stiffness tensor $\mathcal{C}$, has only 21 independent entries (see Chap. 3) and its Voigt representation is defined as a matrix $\mathcal{C}_V \in \text{symm}(\mathbb{R}^{6 \times 6})$ with entries given according to the index map φ as

$$(\mathcal{C}_V)_{\varphi(i,j)\varphi(k,l)} = C_{ijkl} \text{ for } i,j,k,l = 1,2,3.$$

$\square$

This definition allows to write Hooke's law for elasticity Eq. (3.11) in terms of vector-matrix multiplications

$$\boldsymbol{\sigma}_V = \mathcal{C}_V \boldsymbol{\epsilon}_V.$$

The elastic energy Eq. (3.9) is conserved in the sense that:

$$f_{\text{elast}} = \frac{1}{2}\boldsymbol{\sigma}\boldsymbol{\epsilon} = \frac{1}{2}\boldsymbol{\sigma}_V\boldsymbol{\epsilon}_V.$$

Remark In the Voigt notation, stresses and strains are treated differently! When symmetric 3×3 matrices are interpreted in the matrix representation of the elastic linear theory, the interpretation as stress or strain has to be known.

As a tensor is defined via its behavior under orthogonal transformations (cp. Def. 2.17), a way to describe the transformation of tensors in terms of the Voigt representations is needed. The general law of tensor transformation is formulated in terms of direction cosines that determine the change of coordinate systems in three-space. Hence, it cannot be expected to find a rotation matrix $R \in SO(6)$ that describes the coordinate changes for σ_V, ϵ_V and $\mathcal{C}_V$. But, following the book of Newnham [188], the transformation of mechanical stress, strain and stiffness can be done directly in the matrix representation by construction of a suitable matrix $A \in GL(6)$.

Theorem B.1 (TRANSFORMATION OF STRESSES AND STRAINS IN THE MATRIX REPRESENTATION OF HOOKE'S LAW) Let $\mathcal{F}$ and $\mathcal{F}'$ be two frames of reference that are centered at the same position $\mathbf{o} \in \mathbb{R}^3$, and that are related by a rotation $R \in SO(3)$. Let σ_V, ϵ_V and $\mathcal{C}_V$ be the matrix representations of stress, strain and the elastic stiffness tensor in $\mathcal{F}$, and σ'_V, ϵ'_V and $\mathcal{C}'_V$ the matrix representations of stress, strain and the elastic stiffness tensor in $\mathcal{F}'$. There exists a matrix $A(R) \in GL(6)$, such that

(i) $\sigma'_V = A\sigma_V$

(ii) $\epsilon'_V = A^{-T}\epsilon_V$

(iii) $\mathcal{C}'_V = A\mathcal{C}_V A^T$

hold.

PROOF. The proof constructs the matrix $A = (a_{mn})$, where the entries a_{mn} will depend on the entries of the rotation matrix R. From the transformation of the tensors σ and ϵ by application of Hooke's law, the form of the transformation for σ_V is shown. This is basically done by comparing coefficients in both the tensor representation and the Voigt representation. The transformation of the strain tensor σ in the tensor representation is (using the Einstein summation convention):

$$\sigma'_{ij} = r_{ik}r_{jl}\sigma_{kl}.$$

As a matrix with coefficients a_{mn} is to be constructed, this has to fulfill

$$(\sigma'_V)_m = a_{mn}(\sigma_V)_n.$$

Written explicitly for component σ_{11} this reads

$$\sigma'_{11} = r_{11}r_{11}\sigma_{11} + r_{11}r_{12}\sigma_{12} + r_{11}r_{13}\sigma_{13} +$$
$$r_{12}r_{11}\sigma_{21} + r_{12}r_{12}\sigma_{22} + r_{12}r_{13}\sigma_{23} +$$

$$r_{13}r_{11}\boldsymbol{\sigma}_{31} + r_{13}r_{12}\boldsymbol{\sigma}_{32} + r_{13}r_{13}\boldsymbol{\sigma}_{33}$$

and

$$(\boldsymbol{\sigma}'_V)_1 = a_{11}(\boldsymbol{\sigma}_V)_1 + a_{12}(\boldsymbol{\sigma}_V)_2 + a_{13}(\boldsymbol{\sigma}_V)_3 +$$
$$a_{14}(\boldsymbol{\sigma}_V)_4 + a_{15}(\boldsymbol{\sigma}_V)_5 + a_{16}(\boldsymbol{\sigma}_V)_6.$$

Comparison of coefficients leads to the definitions of the a_{mn} as

$$a_{11} = r_{11}^2, \; a_{12} = r_{12}^2, \; a_{13} = r_{13}^2,$$
$$a_{14} = 2r_{12}r_{13}, \; a_{15} = 2r_{13}r_{11}, \; a_{14} = 2r_{12}r_{12}.$$

Applying the same method to the other five independent components of $\boldsymbol{\sigma}$ results in the following matrix (cp. [55]):

$$A = (a_{ij}) =$$

$$\begin{pmatrix} r_{11}^2 & r_{12}^2 & r_{13}^2 & 2r_{12}r_{13} & 2r_{13}r_{11} & 2r_{11}r_{12} \\ r_{21}^2 & r_{22}^2 & r_{23}^2 & 2r_{22}r_{23} & 2r_{23}r_{21} & 2r_{21}r_{22} \\ r_{31}^2 & r_{32}^2 & r_{33}^2 & 2r_{32}r_{33} & 2r_{33}r_{31} & 2r_{31}r_{32} \\ r_{21}r_{31} & r_{22}r_{32} & r_{23}r_{33} & r_{22}r_{33} + r_{23}r_{32} & r_{21}r_{33} + r_{23}r_{31} & r_{22}r_{31} + r_{21}r_{32} \\ r_{31}r_{11} & r_{32}r_{12} & r_{33}r_{13} & r_{12}r_{33} + r_{13}r_{32} & r_{13}r_{31} + r_{11}r_{33} & r_{11}r_{32} + r_{12}r_{31} \\ r_{11}r_{21} & r_{12}r_{22} & r_{13}r_{23} & r_{12}r_{23} + r_{13}r_{22} & r_{13}r_{21} + r_{11}r_{23} & r_{11}r_{22} + r_{12}r_{21} \end{pmatrix}.$$

This proofs $\boldsymbol{\sigma}'_V = A\boldsymbol{\sigma}_V$. $\diagup$
It can be seen that A is invertible with inverse given by (cp. [55])

$$A^{-1} =$$

$$\begin{pmatrix} r_{11}^2 & r_{21}^2 & r_{31}^2 & 2r_{21}r_{31} & 2r_{31}r_{11} & 2r_{11}r_{21} \\ r_{12}^2 & r_{22}^2 & r_{32}^2 & 2r_{22}r_{32} & 2r_{32}r_{12} & 2r_{12}r_{22} \\ r_{13}^2 & r_{23}^2 & r_{33}^2 & 2r_{23}r_{33} & 2r_{33}r_{13} & 2r_{13}r_{23} \\ r_{12}r_{13} & r_{22}r_{23} & r_{32}r_{33} & r_{22}r_{33} + r_{32}r_{23} & r_{12}r_{33} + r_{32}r_{13} & r_{22}r_{13} + r_{12}r_{23} \\ r_{13}r_{11} & r_{23}r_{21} & r_{33}r_{31} & r_{21}r_{33} + r_{31}r_{23} & r_{31}r_{13} + r_{11}r_{33} & r_{11}r_{23} + r_{21}r_{13} \\ r_{11}r_{12} & r_{21}r_{22} & r_{31}r_{32} & r_{21}r_{32} + r_{31}r_{22} & r_{31}r_{12} + r_{11}r_{32} & r_{11}r_{22} + r_{21}r_{12} \end{pmatrix}.$$

The equality $AA^{-1} = \mathbf{I} = A^{-1}A$ follows from the orthogonality of R, as the entry $(AA^{-1})_{ij}$ is related to the (square) of dot products of rows and columns of R. E.g., for $i, j \in \{1, 2, 3\}$

$$(AA^{-1})_{ij} = (R_{i,\cdot} \cdot R_{\cdot,j})^2 = \begin{cases} 0, & \text{if } i \neq j \\ 1^2 = 1, & \text{if } i = j \end{cases}$$

holds ($R_{i,\cdot}$ and $R_{\cdot,j}$ refer to i-th row and j-th column of R).

The product of physical strain and stress is the mechanical energy density, a scalar quantity and therefore a first rank tensor which is invariant under all orthogonal transformations (cp. Eq. (3.9) and Def. 2.17). Writing (in Einstein's notation)

$$\begin{aligned} W &= \epsilon_{ij} \cdot \sigma_{ij} = (\epsilon_V)_i (\sigma_V)_i \\ &= (\epsilon_V)_1 (\sigma_V)_1 + (\epsilon_V)_2 (\sigma_V)_2 + (\epsilon_V)_3 (\sigma_V)_3 + \\ &\quad (\epsilon_V)_4 (\sigma_V)_4 + (\epsilon_V)_5 (\sigma_V)_5 + (\epsilon_V)_6 (\sigma_V)_6 \\ &= (\epsilon_V)^T (\sigma_V) \end{aligned}$$

and exploiting the invariance of W under orthogonal transformations (i.e. $W = W'$), one gets

$$(\epsilon_V')^T (\sigma_V') = W' = W = (\epsilon_V)^T (\sigma_V) = (\epsilon_V)^T A^{-1} A (\sigma_V)$$
$$= (\epsilon_V)^T A^{-1} (\sigma_V'),$$

hence

$$(\epsilon_V')^T = (\epsilon_V)^T A^{-1},$$

or

$$\epsilon_V' = ((\epsilon_V')^T)^T = ((\epsilon_V)^T A^{-1})^T = A^{-T} \epsilon_V,$$

what is the second proposition. $\diagup$

The last equation is equivalent to

$$(A^T \epsilon_V') = (\epsilon_V).$$

From this follows with Hooke's law:

$$C_V' \epsilon_V' = \sigma_V' = A \sigma_V = A(C_V \epsilon_V) = A(C_V A^T \epsilon_V') = (A C_V A^T) \epsilon_V',$$

such that

$$C_V' = (A C_V A^T) \qquad \text{or} \qquad C_V = (A^{-1} C_V' A^{-T}),$$

what shows the third proposition.   /

So, all three proposition are proven.   □

Remark In the numerical implementation of linear elastic effects, the Voigt notation is used for the quantities elastic stress, strain stiffness and compliance. To account for differently oriented phases (grains, martensitic variants, ...), it suffices to compute the matrix $A = A(R) \in GL(6)$ for each phase with respect to the reference coordinate system O_Ω. All needed information is then available from matrix-vector and matrix-matrix operations. This way, computations are much more efficient than using a direct implementation of the tensor formalism.

B.2 Elastic stiffness tensors in Voigt notation

As shown in the last section, the elastic stiffness tensors can be represented by a symmetric 6×6 matrix. This matrix reflects the crystal symmetries the material under consideration provides (cp. the Sec. 2.3). This section lists the most important matrices for stiffness tensor, shows their non-zero components and their interdependencies of the components. These and matrices for other crystal symmetries can e.g. be found in the book of Newnham [55].

Triclinic symmetry The most general case of the elastic stiffness tensor has 21 independent coefficients and occurs when no crystal symmetries are present (cp. Tab. 2.1). This elastic stiffness tensor has the form

$$
\mathcal{C}_{\text{triclinic}} =
\begin{pmatrix}
c_{11} & c_{12} & c_{13} & c_{14} & c_{15} & c_{16} \\
c_{12} & c_{22} & c_{23} & c_{24} & c_{25} & c_{26} \\
c_{13} & c_{23} & c_{33} & c_{34} & c_{35} & c_{36} \\
c_{14} & c_{24} & c_{34} & c_{44} & c_{45} & c_{46} \\
c_{15} & c_{25} & c_{35} & c_{45} & c_{55} & c_{56} \\
c_{16} & c_{26} & c_{36} & c_{46} & c_{56} & c_{66}
\end{pmatrix} .
$$

Tetragonal symmetry Tetragonal symmetry exhibit e.g. the martensitic variants in 5M modulated Ni_2MnGa that occur in the simulations of Chap. 10. The elastic stiffness tensor in the general tetragonal case is of the form

$$\mathcal{C}_{\text{tet}_1} = \begin{pmatrix} c_{11} & c_{12} & c_{13} & 0 & 0 & c_{16} \\ c_{12} & c_{11} & c_{13} & 0 & 0 & -c_{16} \\ c_{13} & c_{23} & c_{33} & 0 & 0 & 0 \\ 0 & 0 & 0 & c_{44} & 0 & 0 \\ 0 & 0 & 0 & 0 & c_{44} & 0 \\ c_{16} & -c_{16} & 0 & 0 & 0 & c_{66} \end{pmatrix}$$

with seven independent entries. The special case when $c_{16} = 0$ reduces the number of independent parameters to six. This reduced form is sufficient to describe the symmetries of tetragonal variants in Ni_2MnGa:

$$\mathcal{C}_{\text{tet}} = \begin{pmatrix} c_{11} & c_{12} & c_{13} & 0 & 0 & 0 \\ c_{12} & c_{11} & c_{13} & 0 & 0 & 0 \\ c_{13} & c_{23} & c_{33} & 0 & 0 & 0 \\ 0 & 0 & 0 & c_{44} & 0 & 0 \\ 0 & 0 & 0 & 0 & c_{44} & 0 \\ 0 & 0 & 0 & 0 & 0 & c_{66} \end{pmatrix}$$

Cubic symmetry The most often assumed symmetry in this work is the case of cubic elastic symmetry. it is reflected by the matrix

$$\mathcal{C}_{\text{cub}} = \begin{pmatrix} c_{11} & c_{12} & c_{12} & 0 & 0 & 0 \\ c_{12} & c_{11} & c_{12} & 0 & 0 & 0 \\ c_{12} & c_{12} & c_{11} & 0 & 0 & 0 \\ 0 & 0 & 0 & c_{44} & 0 & 0 \\ 0 & 0 & 0 & 0 & c_{44} & 0 \\ 0 & 0 & 0 & 0 & 0 & c_{44} \end{pmatrix},$$

where the following relations hold to the so called *Lamé constants* (cp. [43])

$$c_{11} = \lambda + 2\mu + \mu',$$
$$c_{12} = \lambda,$$
$$c_{44} = \mu.$$

So, $\mathcal{C}_{\mathrm{cub}}$ has three independent coefficients. Assuming elastic isotropy, the Lamé constant μ' vanishes, and two independent parameters remain. Thus, in the isotropic case

$$c_{44} = \frac{1}{2}\left(c_{11} - c_{12}\right).$$

B.3 Update scheme for the displacement field

In Sec. 7.3.2, a SOR algorithm is presented that is used to solve the mechanical equilibrium Eq. (7.6). Only the update scheme for the displacement field component $\mathbf{u}_1$ in a cell (i, j, k) in the computation domain is discussed there extensively, as the the update schemes of the other two components of the displacement field are given by application of a transposition[1] of the indexes in the scheme for $\mathbf{u}_1$. The table Tab. 7.1 shows the direct comparison of coefficients and their explicit 'renaming'. For a compact closed formulation of this fact, the tensor notation and the correspondence to the matrix representation of Hooke's law is exploited. If the update scheme for $\mathbf{u}_1$ is defined as a reference, the schemes for the updates of $\mathbf{u}_2$ is gained by switching x_2 and x_3, and the scheme for $\mathbf{u}_3$ by switching x_1 and x_3 and all corresponding indexes. Defining the transpositions $\tau_1 = \mathrm{id}$, $\tau_2 = (2,3)$ and $\tau_3 = (1,3)$, and using the index map φ defined in the beginning of Sec. B.1, the scheme for the update of component u_m ($m = 1, 2, 3$), applied to the tensor notation, reads

$$c_{ijkl} \longmapsto c_{\varphi(\tau_m(i),\tau_m(j))\varphi(\tau_m(k),\tau_m(l))},$$

$$\epsilon_{ij} \longmapsto \epsilon_{\varphi(\tau_m(i),\tau_m(j))}$$

and

$$\sigma_{ij} \longmapsto \sigma_{\varphi(\tau_m(i),\tau_m(j))}$$

for all $i, j, k, l = 1, 2, 3$.

---

[1] A *transposition* is a *permutation* (that is a bijective map of a set containing n elements) that exactly switches two elements. See e.g. the book of Beutelspacher [185] for a proper definition.

Bibliography

[1] B. Nestler, H. Garcke, and B. Stinner. Multicomponent alloy solidification: Phase-field modelling and simulations. *Phys. Rev. E*, 71:041609, 2005.

[2] F. Wendler, J.K. Becker, B. Nestler, P. Bons, and N.P. Walte. Phase-field simulations of partial melts in geological materials. *Comput. Geosci.*, 35:1907 – 1916, 2009.

[3] F. Wendler, C. Mennerich, and B. Nestler. A phase-field model for polycrystalline thin film growth. *J. Cryst. Growth*, 327:189 – 201, 2011.

[4] C. Mennerich, F. Wendler, M. Jainta, and B. Nestler. Phase-field modelling of twin boundary motion in magnetic shape memory alloys. In P. Gumbsch and E. van der Giessen, editors, *MMM 2010 Multiscale Materials Modeling - Conference Proceedings*, 2010.

[5] C. Mennerich, F. Wendler, M. Jainta, and B. Nestler. A Phase-field model for the Magnetic Shape Memory Effect. *Arch. Mech.*, 63(5-6):1845–1875, 2011.

[6] B. Kiefer, T. Bartel, and A. Menzel. A variational approach towards the modeling of magnetic field-induced strains in magnetic shape memory alloys. *J. Mech. Phys. Soldis*, 60:1179 – 1200, 2012.

[7] B. Kiefer, T. Bartel, and A. Menzel. Implementation of numerical integration schemes for the simulation of magnetic sma constitutive response. *Smart Mater. Struct.*, 21:094007, 2012.

[8] C. Mennerich, F. Wendler, M. Jainta, and B. Nestler. Rearrangement of martensitic variants in Ni_2MnGa studied with the phase-field method. *Europ. Phys. J. B*, in press.

[9] A. Braides. A handbook of Γ-convergence. The script is available at the authors homepage: http://axp.mat.uniroma2.it/~braides/home_cv.html.

[10] K. Bhattacharya. *Microstructure of Martensite - Why it forms and how it gives rise to the Shape-Memory Effect.* Oxford University Press, 2003.

[11] A. DeSimone and R. D. James. A constrained theory of magnetoelasticity. *J. Mech. Phys. Solids*, 50:283 – 320, 2002.

[12] D. Lewis and N. Nigam. Geometric integration on spheres and some interesting applications. *J. Comput. Appl. Math.*, 151:141–170, 2003.

[13] C. J. Bradley and A. P. Cracknell. *The Mathematical Theory of Symmetry in Solids: Representation Theory for Point Groups and Space Groups.* Oxford University Press, 2010.

[14] H. Kurzweil and B. Stellmacher. *The Theory of Finite Groups.* Springer, 2004.

[15] O. Deiser. *Einführung in die Mengenlehre.* Springer, 2004.

[16] S. Bosch. *Lineare Algebra.* Springer, 2001.

[17] W. Borchert-Ott. *Kristallographie - Eine Einführung für Naturwissenschaftler.* Springer, 2009.

[18] H. Schade and K. Neemann. *Tensoranalysis.* Walter de Gruyter, 1989.

[19] J. D. Jackson. *Classical Electrodynamics.* John Wiley + Sons, 1999.

[20] P. McCord Morse and H. Feshbach. *Methods of Theoretical Physics, Part I.* McGraw-Hill Science/Engineering/Math, 1953.

[21] Y.F. Gui and W.B. Dou. A rigorous and completed statement on Helmholtz Theorem. *Prog. Electromagn. Res.*, 69:287 – 304, 2007.

[22] T. Hahn, editor. *International Tables for Crystallography, Volume A: Space Group Symmetry.* Springer, 5th edition, 2002.

[23] D. Schwarzenbach. *Kristallographie.* Springer, 2001.

[24] A. Schoenflies. *Krystallsysteme und Krystallstructur.* Springer, Repr. d. Ausg. Leipzig, Teubner, 1891 / Xii, 638 S. : graph. Darst. edition, 1984.

[25] H. S. M. Coxeter. *Regular Polytopes.* Dover Publications, 1973.

[26] L. Bieberbach. Über die Bewegungsgruppen der Euklidischen Räume. (Zweite Abhandlung.) Die Gruppen mit einem endlichen Fundamentalbereich. *Mathematische Annalen*, 72:400 – 412, 1912.

[27] L. A. Skornjakov. *Elemente der Verbandstheorie*. Akademie-Verlag, 1973.

[28] R. N. Bracewell. *The Fourier transform and its applications*. McGraw-Hill Kogakusha, 1978.

[29] K. Königsberger. *Analysis 2*. Springer, 2002.

[30] A. Iserles, H. Z. Munthe-Kaas, S. P. Nørsett, and A. Zanna. *Lie-group methods*. Cambridge University Press, 1999.

[31] A. Iserles. Lie-group methods in geometric numerical integration. Talk, available at: http://na.math.kit.edu/marlis/download/meetings/08Oberwolfach/Lie_group-print.pdf.

[32] J. Hilgert and K. H. Neeb. *Lie-Gruppen und Lie-Algebren*. Vieweg, 1991.

[33] W. M. Lai, D. Rubin, and E. Krempl. *Introduction to Continuum Mechanics*. Elsevier, 4th edition, 2010.

[34] W. Jaunzemis. *Continuum Mechanics*. Collier Macmillan, 1967.

[35] R. Phillips. *Crystals, Defects and Microstructure: Modeling across Scales*. Cambridge University Press, 2000.

[36] M. Šilhavý. *The Mechanics and Thermodynamics of Continuous Media*. Springer, 1997.

[37] E. K. H. Salje. *Phase Transitions in Ferroelastic and Co-elastic Crystals*. Oxford University Press, 1993.

[38] A. Bertram and B. Svendsen. On material objectivity and reduced constitutive equations. *Arch. Mech.*, 53:653 – 675, 2001.

[39] B. Svendsen and A. Bertram. On frame-indifference and form-invariance in constitutive theory. *Acta Mech.*, 132:195 – 207, 1999.

[40] T. W. B. Kibble and F. H. Berkshire. *Classical Mechanics*. World Scientific Publishing Company, 5th edition, 2004.

[41] S. Bharatha and M. Levinson. On physically nonlinear elasticity. *J. Elasticity*, 7(3):287 – 304, 1977.

[42] M. Arndt, M. Griebel, V. Novák, T. Roubíček, and P. Šittner. Martensitic transformation in NiMnGa single crystals: Numerical simulation and experiments. *Int. J. Plasticity*, 22:1943 – 1961, 2006.

[43] D. Gross and T. Seelig. *Bruchmechanik: Mit einer Einführung in die Mikromechanik.* Springer, 2011.

[44] A. Hubert and R. Schäfer. *Magnetic Domains.* Springer, 1998.

[45] A. Aharoni. *Introduction to the theory of Ferromagnetism.* Oxford Science Publication, 2000.

[46] R. C. O'Handley. *Modern Magnetic Materials.* John Wiley & Sons, 2000.

[47] H. B. Callen. *Thermodynamics and an Introduction to Thermo-statistics.* Wiley, 1985.

[48] M. Plischke and B. Bergersen. *Equilibrium Statistical Physics.* World Scientific Publishing Company, 1994.

[49] P. C. Scholten. Which SI? *J. Magn. Magn. Mater.*, 149:59, 1995.

[50] J. Stöhr and H. C. Siegmann. *Magnetism - From Fundamentals to Nanoscale Dynamics.* Springer, 2006.

[51] W. F. Brown Jr. *Micromagnetics.* Interscience Publisher, 1963.

[52] J. E. Miltat and M. J. Donahue. Numerical micromagnetics: Finite difference methods. In H. Kronmüller and S. Parkin, editors, *Handbook of Magnetism and Advanced Magnetic Materials.* John Wiley and Sons Ltd., 2007.

[53] J. X. Zhang and L. Q. Chen. Phase-field model for ferromagnetic shape-memory alloys. *Phil. Mag. Lett.*, 85:531 – 541, 2005.

[54] H. Kronmüller. General micromagnetic theory. In H. Kronmüller and S. Parkin, editors, *Handbook of Magnetism and Advanced Magnetic Materials.* John Wiley and Sons Ltd., 2007.

[55] R. E. Newnham. *Properties of materials: anisotropy, symmetry, structure.* Oxford University Press, 2005.

[56] I. Cimrák. A survey on the numerics and computations for the Landau-Lifshitz equation of micromagnetism. *Arch. Comput. Methods. Eng.*, 15:277 – 309, 2008.

[57] O. Heczko. Magnetic shape memory effect and magnetization reversal. *J. Magn. Magn. Mater.*, 290 – 291:787 – 794, 2005.

[58] Y. M. Jin. Domain microstructure in magnetic shape memory alloys: Phase-field model and simulation. *Acta. Mater.*, 57:2488 – 2495, 2009.

[59] K. Ullakko, J. K. Huang, C. Kantner, R. C. O'Handley, and V. V. Kokorin. Large magnetic-field-induced strains in Ni_2MnGa single crystals. *Appl. Phys. Lett.*, 69:1966 – 1968, 1996.

[60] P. Entel, V. D. Buchelnikov, M. E. Gruner, A. Hucht, V. V. Khovailo, S. K., and N. A. T. Zayak. Shape memory alloys: A summary of recent achievements. *Mater. Sci. Forum*, 583:21 – 41, 2008.

[61] J. Slutsker and A. L. Roytburd. Deformation of adaptive materials. part i. constrained deformation of polydomain crystals. *J. Mech. Phys. Solids*, 47:2299 – 2329, 1999.

[62] J. Slutsker and A. L. Roytburd. Deformation of adaptive materials. part ii. adaptive composite. *J. Mech. Phys. Solids*, 47:2331 – 2349, 1999.

[63] J. Slutsker and A. L. Roytburd. Deformation of adaptive materials. part iii: Deformation of crystals with polytwin product phases. *J. Mech. Phys. Solids*, 49:1795 – 1822, 2001.

[64] Y. Wang and A.G. Khachaturyan. Multi-scale phase field approach to martensitic transformations. *Mater. Sci. Eng. A*, 438 - 440:55 – 63, 2006.

[65] V.I. Levitas, D.-W. Lee, and D.L. Preston. Interface propagation and microstructure evolution in phase field models of stress-induced martensitic phase transformations. *Int. J. Plasticity*, 26:395 – 422, 2010.

[66] J. Kundin, D. Raabe, and H. Emmerich. A phase-field model for incoherent martensitic transformations including plastic accomodation processes in the austenite. *J. Mech. Phys. Solids*, 59:2082 – 2102, 2011.

[67] M. L. Richard, J. Feuchtwanger, S. M. Allen, R. C. O'Handley, P. Lázpita, and J. M. Barandiaran. Martensite transformation in Ni-Mn-Ga ferromagnetic shape-memory alloys. *Metall. Mater. Trans. A*, 38:777–780, 2007.

[68] D.A. Garanin. Fokker-planck and Landau-Lifshitz-Bloch equations for classical ferromagnet. *Phys. Rev. B*, 55:3050 – 3057, 1997.

[69] C. Schieback, D. Hinzke, M. Kläui, U. Nowak, and P. Nielaba. Temperature dependence of the current-induced domain wall motion from a modified Landau-Lifshitz-Bloch. *Phys. Rev. B*, 80:214403, 2009.

[70] K. Otsuka and C. M. Wayman, editors. *Shape Memory Materials*. Cambridge University Press, 1998.

[71] G. Gottstein. *Physikalische Grundlagen der Materialkunde*. Springer, 2007.

[72] V. C. Solomon, M. R. McCartney, D. J. Smith, J. Tang, A. E. Berkowitz, and R. C. O'Handley. Magnetic domain configurations in spark-eroded ferromagnetic shape memory Ni-Mn-Ga particles. *Appl. Phys. Lett.*, 86:192503-1 – 192503-3, 2005.

[73] D.C. Lagoudas, editor. *Introduction to Shape Memory Alloys*, pages 1–51. Springer, 2008.

[74] D. J. Seol, S. Y. Hu, Y. L. Li, J. Shen, , K. H. Oh, and L. Q. Chen. Three-dimensional Phase-Field Modeling of Spinodal Decomposition in Constrained Films. *Met. Mater. Int.*, 9(1):61 – 66, 2003.

[75] L. Q. Chen. Phase-field models for microstructure evolution. *Annu. Rev. Mater. Res.*, 32:113 – 140, 2002.

[76] R .S. Qin and H. K. Bhadeshia. Phase-field models for microstructure evolution. *Annu. Rev. Mater. Res.*, 10:803 – 811, 2010.

[77] Nele Moleans, Bart Blanpain, and Patrick Wollants. An introduction to phase-field modeling of microstructure evolution. *Calphad*, 32:268 – 294, 2008.

[78] A. Choudhury and B. Nestler. Grand-potential formulation for multicomponent phase transformations combined with thin-interface asymptotics of the double-obstacle potential. *Phys. Rev. E*, 85:021602-1, 2012.

[79] C. Eck, H. Garcke, and P. Knabbner. *Mathematische Modellierung*. Springer, 2008.

[80] J.S. Rowlinson. Translation of j. d. van der waals the thermodynamic theory of capillarity under the hypothesis of a continuous variation of density. *J. Stat. Phys.*, 28:258 – 267, 1979.

[81] John W. Cahn and John E. Hilliard. Free energy of a nonuniform system. I. Interfacial free energy. *J. Chem. Phys.*, 28:258 – 267, 1958.

[82] J.S. Rowlinson. Theory of the dispersion of magnetic permeability in ferromagnetic bodies. *Phys. Zeitsch. der Sow.*, 8:153 – 169, 1935.

[83] J.S. Rowlinson. A microscopic theory for domain wall motion and its experimental verification in Fe-Al alloy domain growth kinetics. *J. de Phsy.*, 12:51 – 54, 1977.

[84] D. Raabe, F. Roters, F. Barlat, and L. Q. Chen, editors. *Introduction to the phase-field method of microstructure evolution*, pages 37 – 56. Wiley, 2004.

[85] H. Garcke, B. Nestler, and B. Stoth. On anisotropic order parameter models for multi-phase systems and their sharp interface limits. *Physica D*, 115:87 – 108, 1998.

[86] H. Garcke, B. Nestler, and B. Stinner. A diffuse interface model for alloys with multiple components and phases. *SIAM J. Appl. Math.*, 64:775–799, 2004.

[87] O. Heczko. Determination of ordinary magnetostriction in Ni-Mn-Ga magnetic shape memory alloys. *J. Magn. Magn. Mater.*, 290-291:846 – 849, 2004.

[88] J. X. Zhang and L. Q. Chen. Phase-field microelasticity theory and micromagnetic simulations of domain structures in giant magnetostrictive materials. *Acta. Mater.*, 53:2845 – 2855, 2005.

[89] A. Artemev, Y. Jin, and A.G. Khachaturyan. Three-dimensional phase-field model of proper martensitic transformation. *Acta Mater.*, 49:1165 – 1177, 2001.

[90] P. P. Wu, X. Q. Ma, J. X. Zhang, and L. Q. Chen. Phase-field simulations of stress-strain behaviour in ferromagnetic shape memory alloy Ni_2MnGa. *J. Appl. Phys.*, 104:073906, 2008.

[91] L. D. Landau and E. M. Lifshitz. *Statistical Physics*. Oxford: Pergamon Press, 1980.

[92] A. P. Cracknell. Group theory in solid-state physics is not dead yet alias some recent developments in the use of group theory in solid-state physics. *Adv. Phys.*, 23:673 – 866, 1974.

[93] E. Salje and V. Devarajan. Phase transitions in systems with strain-induced coupling between two order parameters. *Phase Transitions: A multinational Journal*, 6(3):235 – 247, 1986.

[94] P. P. Wu, X. Q. Ma, J. X. Zhang, and L. Q. Chen. Phase-field simulations of magnetic field-induced strain in Ni_2MnGa ferromagnetic shape memory alloy. *Phil. Mag.*, 91:2112 – 2116, 2011.

[95] A. N. Vasil'ev, A. D. Bozhko, and V. V. Khovailo. Structural and magnetic phase transitions in shape-memory alloys $Ni_{2+x}Mn_{1+x}Ga$. *Phys. Rev. B*, 59(2):1113 – 1120, 1999.

[96] L. J. Li, J. Y. Li, Y. C. Shu, H. Z. Chen, and J. H. Yen. Magnetoelastic domains and magnetic field-induced strains in ferromagnetic shape memory alloys by phase-field simulation. *Appl. Phys. Lett.*, 92:172504-1 – 172504-3, 2008.

[97] C. M. Landis. A continuum thermodynamics formulation for micro-magneto-mechanics with applications to ferromagnetic shape memory alloys. *J. Mech. Phys. Solids*, 56:3059 – 3076, 2008.

[98] B. Nestler. A 3d parallel simulator for crystal growth and solidification in complex alloy systems. *J. Crys. Growth*, 275(1-2):273 – 278, 2005.

[99] R. Spatschek, C. Müller-Gugenberger, E. Brener, and B. Nestler. Phase-field modeling of fracture and stress induced phase transitions. *Phys. Rev. E*, 75:066111, 2007.

[100] M. Jainta. Simulation von Formgedächtnismaterialien mit Hilfe eines Phasenfeldmodells. Bachelor's thesis, Hochschule Karlsruhe - Technik und Wirtschaft, 2009.

[101] H.-R. Schwarz. *Numerische Mathematik*. Teubner Verlag, 1997.

[102] J. Stoer. *Numerische Mathematik 1*. Springer, 1989.

[103] B. Nestler, M. Reichardt, and M. Selzer. Massive multi-phase-field simulations: Methods to compute large grain system. In J. Hirsch, B. Skrotzki, and G. Gottstein, editors, *Proceedings of the 11th International Conference on Aluminium Alloys, 22 - 26 Sept. 2008, Aachen, Germany*, pages 1251–1255. WILEY-VCH Verlag GmbH and Co.KGaA, Weinheim, 2008.

[104] S. G. Kim, D. I. Kim, W. T. Kim, and Y. B. Park. Computer simulation of two-dimensional and three-dimensional ideal grain growth. *Phys. Rev. E*, 74:061605–1–061605–14, 2006.

[105] M. Griebel, T. Dornseifer, and T. Neunhoeffer. *Numerische Simulation in der Strömungsmechanik*. vieweg, 1995.

[106] E. H. Saenger. *Wave Propagation in Fractured Media: Theory and Applications of the Rotated Staggered Finite-Difference Grid*. PhD thesis, Universität Karlsruhe (TH), 2000.

[107] J.H. Hattel and P.N. Hansen. A control volume-based finite difference method for solving the equilibrium equations in terms of displacements. *Appl. Math. Modelling*, 19, 1995.

[108] S. Nemat-Nasser. *Micromechanics: Overall Properties of Heterogeneous Materials*. Elsevier Ltd, 1998.

[109] A. Fröhlich. *Mikromechanisches Modell zur Ermittlung effektiver Materialeigenschaften von piezoelektrischen Polykristallen*. PhD thesis, Universität Karlsruhe (TH), 2001.

[110] A. G. Khachaturyan. *Theory of structural transformations in solids*. Wiley, 1983.

[111] I. Steinbach and M. Apel. Multi phase field model for solid state transformation with elastic strain. *Physica D*, 217:153 – 160, 2006.

[112] D. Lewis and N. Nigam. A geometric integration algorithm with applications to micromagnetics. *Tech. Rep. 1721, IMA Preprint Series*, 2000.

[113] X.-P. Wang, C. J. García-Cervera, and W. E. A Gauss-Seidel Projection Method for Micromagnetics Simulations. *J. Comp. Phys.*, 171:357 – 372, 2001.

[114] K.M. Lebecki, M.J. Donahue, and M.W. Gutowski. Periodic boundary conditions for demagnetization interactions in micromagnetic simulations. *J. Phys. D: Appl. Phys.*, 41:175005, 2008.

[115] W. Wang, C. Mu, B. Zhang, Q. Liu, J. Wang, and D. Xue. Two-dimensional periodic boundary conditions for demagnetization interactions in micromagnetics. *Comp. Mater. Science*, 49:84 – 87, 2010.

[116] A. J. Newell, W. Williams, and D. J. Dunlop. A generalization of the demagnetizing tensor for nonuniform magnetization. *J. Geophys. Res.*, 98:9551 – 9555, 1993.

[117] J. Caro, M. Noack, P. Kölsch, and R. Schäfer. Zeolite membranes - state of their development and perspective. *Microporous Mesoporous Mater.*, 38:3–24, 2000.

[118] G. Xomeritakis, A. Gouzinis, S. Nair, T. Okubo, M. He, R. M. Overney, and M. Tsapatsis. Growth, microstructure, and permeation properties of supported zeolite (MFI) films and membranes prepared by secondary growth. *Chem. Eng. Sci.*, 54:3521–3531, 1999.

[119] J. Hedlund and F. Jareman. Texture of MFI films grown from seeds. *Curr. Opin. Colloid Interface Sci.*, 10:226–232, 2005.

[120] A. Gouzinis and M. Tsapatsis. On the preferred orientation and microstructural manipulation of molecular sieve films prepared by secondary growth. *Chem. Mater.*, 10:2497–2504, 1998.

[121] J. Kärger. Random walk through two-channel networks: a simple means to correlate the coefficients of anisotropic diffusion in ZSM-5 type zeolites. *J. Phys. Chem.*, 95:5558–5560, 1991.

[122] I. Díaz, E. Kokkoli, O. Terasaki, and M. Tsapatsis. Surface structure of Zeolite (MFI) crystals. *Chem. Mater.*, 16:5226–5232, 2004.

[123] J. K. Becker, P. D. Bons, and M. W. Jesell. A new front-tracking method to model anisotropic grain and phase boundary motion in rocks. *Comput. Geosci.*, 34:201–212, 2008.

[124] W. J. Boettinger, J. A. Warren, C. Beckermann, and A. Karma. Phase-field simulation of solidification. *Annu. Rev. Mater. Res.*, 32:163–194, 2002.

[125] R. Kobayashi, J. A. Warren, and W. C. Carter. Vector-valued phase-field model for crystallization and grain boundary formation. *Physica D*, 119:415 – 423, 1998.

[126] J. A. Warren, R. Kobayashi, A. E. Lobkovsky, and W. C. Carter. Extending phase-field models of solidification to polycrystalline materials. *Acta Mater.*, 51:6035–6058, 2003.

[127] L. Gránásy, T. Pusztai, T. Börzsönyi, J. A. Warren, and J. F. Douglas. A general mechanism of polycrystalline growth. *Nat. Mater.*, 3:645–650, 2004.

[128] P. Chen, Y. L. Tsai, and C. W. Lan. Phase field modelling of growth competition of silicon grains. *Acta Mater.*, 56:4114–4122, 2008.

[129] J. J. Eggleston, G. B. McFadden, and P. W. Voorhees. A phase-field model for highly anisotropic interfacial anisotropy. *Physica D*, 150:91 – 103, 2001.

[130] E. Yokoyama and R. F. Sekerka. A numerical study of the combined effect of anisotropic surface tension and interface kinetics on pattern formation during the growth of two dimensional crystals. *J. Cryst. Growth*, 125:389 – 403, 1992.

[131] T. Uehara and R. F. Sekerka. Phase field simulations of faceted growth for strong anisotropy of kinetic coefficient. *J. Cryst. Growth*, 254:251–261, 2003.

[132] R. F. Sekerka. Equilibrium and growth shapes of crystals: how do they differ and why should we care? *Cryst. Res. Technol.*, 40(4-5):291–306, 2005.

[133] H. Garcke, B. Stoth, and B. Nestler. Anisotropy in multi-phase systems: a phase field approach. *Interfaces Free Bound.*, 1:175–198, 1999.

[134] E. Galli, G. Vezzalini, S. Quartieri, A. Alberti, and M. Franzini. Mutinaite, a new zeolite from Antarctica: The natural counterpart of ZSM-5. *Zeolites*, 19:318–322, 1997.

[135] D. L. Dorset. Electron crystallography of zeolites. 1. projected crystal structures of ZSM-5 and ZSM-11. *Z. Kristallogr.*, 218:458–465, 2003.

[136] W. Schmidt, U. Wilczok, C. Weidenthaler, O Medenbach, R. Goddard, G. Buth, and A. Cepak. Preparation and morphology of pyramidal MFI single-crystal segments. *J. Phys. Chem. B*, 111:13538–13543, 2007.

[137] A.-J. Bons and P. D. Bons. The development of oblique prefered orientations in zeolite films and membranes. *Microporous Mesoporous Mater.*, 62:9–16, 2003.

[138] X. Chen, W. Yan, X. Cao, J. Yu, and R. Xu. Fabrication of silicalite-1 crystals with tunable aspect ratios by microwave-assisted solvothermal synthesis. *Microporous Mesoporous Mater.*, 119:217–222, 2009.

[139] C. B. de Gruyter, J. P. Verduijn, J. Y. Koo, S. B. Rice, and M. M. J. Treacy. A transmission electron microscopy study of grain boundaries in zeolite l. *Ultramicroscopy*, 34:102–107, 1990.

[140] F. Wendler, J. K. Becker, B. Nestler, P. D. Bons, and N. P. Walte. Phase-field simulations of partial melts in geological materials. *Comput. Geosci.*, 35:1907–1916, 2009.

[141] A. Navrotsky. Energetic clues to pathways to biomineralization: Precursors, clusters, and nanoparticles. *Proc. Nat. Acad. Sci. U.S.A.*, 101:12096–12101, 2004.

[142] V. Nikolakis, E. Kokkoli, M. Tirrell, M. Tsapatsis, and D. G. Vlachos. Zeolite growth by addition of subcolloidal particles: Modeling and experimental validation. *Chem. Mater.*, 12:845–853, 2000.

[143] S. Yang and A. Navrotsky. In situ calorimetric study of the growth of silica TPA-MFI crystals from an initially clear solution. *Chem. Mater.*, 14:2803–2811, 2002.

[144] F. Andreola, M. Romagnoli, C. Siligardi, T. Manfredini, C. Ferone, and M. Pansini. Densification and crystallization of Ba-exchanged zeolite A powders. *Ceram. Int.*, 34:543–549, 2008.

[145] J. M. Thijssen, H. J. F. Knops, and A. J. Dammers. Dynamic scaling in polycrystalline growth. *Phys. Rev. B*, 45:8650–8656, 1992.

[146] J. M. Thijssen. Simulations of polycrystalline growth in 2+1 dimensions. *Phys. Rev. B*, 51:1985–1988, 1995.

[147] A. Morawiec. On equilibrium conditions at junctions of anisotropic interfaces. *J. Mater. Sci.*, 40:2803–2806, 2005.

[148] J. A. Thornton. High rate thick film growth. *Ann. Rev. Mater. Sci.*, 7:239–260, 1977.

[149] A. Van der Drift. *Philips Res. Rep.*, 22:267–288, 1967.

[150] G. Carter. Preferred orientation in thin film growth - the survival of the fastest. *Vacuum*, 56:87–93, 2000.

[151] O. Nilsen, O. B. Karlsen, A. Kjekshus, and H. Fjellvøg. Simulation of growth dynamics in atomic layer deposition. part ii. polycrystalline films from cubic crystallites. *Thin Solid Films*, 515:4538–4549, 2007.

[152] O. Nilsen, O. B. Karlsen, A. Kjekshus, and H. Fjellvøg. Simulation of growth dynamics in atomic layer deposition. part iii. polycrystalline films from tetragonal crystallites. *Thin Solid Films*, 515:4550–4558, 2007.

[153] A. B. Rodriquez-Navarro. Model of texture development in polycrystalline films growing on amorphous substrates with different topographies. *Thin Solid Films*, 389:288–295, 2001.

[154] D. J. Paritosh, D. J. Srolovitz, C. C. Battaile, X. Li, and J. E. Butler. Simulation of faceted film growth in two-dimensions: Microstructure, morphology and texture. *Acta Mater.*, 47:2269–2281, 1999.

[155] P. Smereka, X. Li, G. Russo, and D. J. Srolovitz. Simulations of faceted film growth in three dimensions: microstructure, morphology and texture. *Acta Mater.*, 53:1191–1204, 2005.

[156] C. S. Cundy and P. A. Cox. The hydrothermal synthesis of zeolites: Precursors, intermediates and reaction mechanisms. *Microporous Mesoporous Mater.*, 82:1–78, 2005.

[157] B. Nestler, H. Garcke, and B. Stinner. Multicomponent alloy solidification: phase-field modelling and simulations. *Phys. Rev. E*, 71:041609–1, 2005.

[158] D.C. Lagoudas, editor. *Modelling of Magentic SMAs*, pages 325–393. Springer, 2008.

[159] C. Miehe, B. Kiefer, and D. Rosato. An incremental variational formulation of dissipative magnetostriction at the macroscopic continuum level. *Int. J. Solids Struct.*, 48:1846 – 1866, 2011.

[160] S. Conti, M. Lenz, and M. Rumpf. Modeling and simulation of magnetic shape-memory polymer composites. *J. Mech. Phys. Solids*, 55:1462 – 1486, 2007.

[161] L. J. Li, C.H. Lei, Y. C. Shu, and J. Y. Li. Phase-field simulation of magnetoelastic couplings in ferromagnetic shape memory alloys. *Acta Mater.*, 59:2648 – 2655, 2011.

[162] B. Krevet and M. Kohl. FEM simulation of a Ni-Mn-Ga film bridge actuator. *Physics Procedia*, 10:154 – 161, 2010.

[163] Y. Su and C. M. Landis. Continuum thermodynamics of ferroelectric domain evolution: Theory, finite element implementation, and application to domain wall pinning. *J. Mech. Phys. Solids*, 55:280 – 305, 2007.

[164] D. Schrade, R. Mueller, B.X. Xu, and D. Gross. Domain evolution in ferroelectric materials: A continuum phase field model and finite element implementation. *Comput. Methods Appl. Mech. Eng.*, 196:4365 – 4374, 2007.

[165] R. Tickle and R. D. James. Magnetic and magnetomechanical properties of Ni_2MnGa. *J. Magn. Magn. Mater*, 195:627 – 638, 1999.

[166] O. Heczko and K. Ullakko. Effect of temperature on magnetic properties of Ni-Mn-Ga Magnetic Shape Memory (MSM) alloys. *IEEE Trans. Magn.*, 37(4):2672 – 2674, 2001.

[167] L. Straka, O. Heczko, and H. Hänninen. Activation of magnetic shape memory effect in Ni-Mn-Ga alloys by mechanical and magnetic treatment. *Acta Mater.*, 56:5492 – 5499, 2008.

[168] E. Faran and D. Shilo. Implications of twinning kinetics on the frequency response in NiMnGa actuators. *Appl. Phys. Let.*, 100:151901, 2012.

[169] A. Finel, Y. Le Bouar, A. Gaubert, and U. Salmana. Phase field methods: Microstructures, mechanical properties and complexity. *C. R. Phys.*, 11:245 – 256, 2010.

[170] Y. W. Lai, N. Scheerbaum, D. Hinz, O. Gutfleisch, R. Schäfer, L. Schultz, and J. McCord. Absence of magnetic domain wall motion during magnetic field induced twin boundary motion in bulk magnetic shape memory alloys. *Appl. Phys. Lett.*, 90:192504, 2007.

[171] T. Waitz, T. Antretter, F. D. Fischer, N.K. Simhad, and H. P. Karnthaler. Size effects on the martensitic phase transformation of NiTi nanograins. *J. Mech. Phys. Soldis*, 55:419 – 444, 2007.

[172] M. J. Donahue. Accurate computation of the demagnetization tensor. Talk, available at: http://math.nist.gov/ MDonahue/talks.html/.

[173] E. Warburg. Magnetische Untersuchungen. *Ann. Phys. (Leipzig)*, 13:141 – 164, 1881.

[174] S. Fähler, U. K. Rössler, O. Kastner, J. Eckert, G. Eggeler, H. Emmerich, P. Entel, P. Müller, E. Quandt, and K. Albe. Caloric effects in ferroic materials: New concepts for cooling. *Adv. Eng. Mater.*, 13, 2011.

[175] V. K. Pecharsky and K. A. Gschneidner Jr. Tunable magnetic regenerator alloys with a giant magnetocaloric effect for magnetic refrigeration from 20 to 290 K. *Appl. Phys. Lett.*, 70:3299 – 3301, 1997.

[176] L. Mañosa, D. González-Alonso, A. Planes, E. Bonnot, M. Barrio, J.-L. Tamarit, S. Aksoy, and M. Acet. Giant solid-state barocaloric effect in the Ni-Mn-In magnetic shape-memory alloy. *Nat. Mater.*, 9:478 – 481, 2010.

[177] T. Krenke, E. Duman, M. Acet, E. F. Wassermann, X. Moya, L. Mañosa, and A. Planes. Inverse magnetocaloric effect in ferromagnetic Ni-Mn-Sn alloys. *Nat. Mater.*, 4:450–454, 2005.

[178] J. Marcos, L. Mañosa, A. Planes, F. Casanova, X. Batlle, A. Labarta, and B. Martinez. Magnetic field induced entropy changes in Ni-Mn-Ga alloys. *J. Magn. Magn. Mater.*, 272-276:e1595 – e1596, 2004.

[179] V. K. Pecharsky and K. A. Gschneidner Jr. Magnetocaloric effect and magnetic refrigeration. *J. Magn. Magn. Mater.*, 200:44 – 56, 1999.

[180] K. A. Gschneidner Jr., V. K. Pecharsky, and A. O. Tsokol. Recent developments in magnetocaloric materials. *Rep. Prog. Phys.*, 68:1479 – 1539, 2005.

[181] K. A. Gschneidner Jr. and V. K. Pecharsky. Magnetocaloric materials. *Annu. Rev. Mater. Sci.*, 30:387 – 429, 2000.

[182] N. A. de Õliveira and P. J. von Ranke. Theoretical aspects of the magnetocaloric effect. *Phys. Rep.*, 489:89 – 159, 2010.

[183] H. Garcke, B. Nestler, and B. Stinner. A diffuse interface model for alloys with multiple components and phases. *SIAM J. Appl. Math.*, 64:775–799, 2004.

[184] H. Goldstein, C. P. Poole, Jr, and J. L. Safko. *Klassische Mechanik.* Wiley-VCH, 2006.

[185] A. Beutelspacher. *Lineare Algebra: Eine Einführung in die Wissenschaft der Vektoren, Abbildungen und Matrizen.* Vieweg+Teubner Verlag, 2010.

[186] G. Stoth. *Algebra. Einführung in die Galoistheorie.* deGruyter, 1998.

[187] J. M. Selig. *Geometric Fundamentals of Robotics.* Springer, 2005.

[188] R. E. Newnham. *Properties of Materials: Anisotropy, Symmetry, Structure.* Oxford University Press, 2005.

Index

Band 19   Christian Mennerich
           **Phase-field modeling of multi-domain evolution in ferromagnetic
           shape memory alloys and of polycrystalline thin film growth.** 2013
           ISBN 978-3-7315-0009-4